Insane in Ukraine

William H. Volpé III, PMP

Author's Note

This book consists of details related to the Ukrainian war and the Odesa, Ukraine massacre. It also includes the author's time in Saudi Arabia, Egypt, Bahrain, and Moldova, before finally arriving in South America. The activities and events in this book took place from 2013 through 2015.

INSANE IN UKRAINE

Published by The Volpé Consortium, Inc.

Printed in the United States of America.

For more information, please write:

The Volpé Consortium, Inc.

Email: contact@thevolpeconsortium.com

or call 1 (512) 632-5172

Visit us online at: www.thevolpeconsortium.com

ISBN: 978-1-7332413-1-1

Copyright: BKR621-75

Abstract

In 2013, William H. Volpé III, an international businessman and project manager, that was a former US government contractor under SECRET Clearance, embarked on an adventure that would change his life forever. He was between projects and there was a pending effort in Saudi Arabia for the Ministry of Interior that he was being considered for. During his downtime, he decided to go to Ukraine where he met Natasha and fell in love.

Natasha was a USSR born Ukrainian woman that was formerly a lawyer for the Ukrainian Ministry of Interior and had a son. She wanted to see the world and had already traveled to many places, but there were still cultural issues that surrounded old soviet families engaging with Americans. Ultimately, the two were married in secret on the eve of the Ukraine war and the Odesa massacre, which would prove to be the largest single day of death in Ukraine since the end of WWII.

INSANE IN UKRAINE

Dedication

This book is dedicated to everyone that made it out of Ukraine alive and to the memories of all of those who didn't.

Natasha, you will always be one of the great loves of my life that changed my path forever.

Volodia, I hope that you always remain safe and never lose your innocence.

Brenda, thank you for the time you put into editing this project and discussing with me the many directions that it could go.

To all of the brothers that never forgot about me or where I was and what was going on at the time, you got me through. Thank you.

Table of Contents

Introduction

There's no celebrity introduction here. There are no kudos. The reason is this is another one of my "different" books. We all start out with this "plan" in mind, but I think Mike Tyson said it best when he said everyone has a plan until they get hit in the nose. Not sure if the nose was the most accurate part of this analogy. I think this one hit me in the balls. I, like many guys, actually love my nuts, too. It's one of the best relationships I have. However, this trip, this experience, this life, again, got weird(er).

For those of you that don't know me, I am William H. Volpé III (a.k.a. Bill Volpé). It's a mouthful, I know. Some even took to calling me "Wild Bill" Volpé. I was born on Parris Island and grew up military. I was an international project manager that traveled all over the world and started my career as a US government contractor under SECRET Clearance. This book is a true story about how I ended up in the Odesa, Ukraine massacre during the civil war and Russian occupation before finally escaping through Moscow to Ecuador. It will detail the time I spent in Saudi Arabia, Egypt, Moldova, Bahrain, and Ukraine in my typical salty, conversational style since that is just who I am.

My father did three tours in Vietnam and was a Marine Corps sniper and my grandfather was a major on General Bradley's staff when

Bradley commanded all the American D-Day troops in WWII. Grandfather then went on to Korea to work in the Korean Military Advisory Group. On my grandmother's side, we dated back to such people like Daniel Boone. So, it would seem to reason, adventure was in my future. At least, it was in my past and hanging on the family tree.

My biological parents never really seemed to want a child or know what to do with him, so I was raised by my grandparents on my father's side, which did a lot to shape my life including international travel. Of course, I am quick to say that being born Greek – Irish on Hell Island during the July 4th weekend in 1968 in the USA to a drill instructor pretty much means you are fucked at birth. How fucked one never knows until they just ride it in. It reminds me of a line at the end of Boondock Saints when they ask their dad, "How far are we going to go with this?" Their dad replies, "It is not a question of how far, but a question of do you have the power of conviction to go as far as needed." When it came to crazy like a fox stupid, I never seemed to be short of conviction.

I had an international project management training and consulting company, and I, like many others, liked the freedom of being able to use technology to manage projects from anywhere in the world. Not so much because I liked to travel, but more because you never knew when you needed to flee from prosecution. In any event, in 2013, I had some downtime, sort of, where I didn't actually need to be somewhere and I had been growing closer to a beautiful woman who lived in Odesa,

Ukraine. Since everything I had to manage could be managed remotely, and I was greatly missing Natasha, it seemed a great time to head to Odesa, Ukraine for a month or so.

I had found over the years that the 20-30 day mark was about the time people started saying “I haven’t seen Bill for a while. Do you think he is still hungover or actually dead this time?” The “mail people” and “trash people” didn’t seem to care so much. The mail people on the lake I lived on would just let your shit fall out of the box or give it to someone else and the trash people were happy that they didn’t have to do maintenance on their truck from picking up barrels of beer bottles. So, basically, the 20-30 day mark was when my friends got sober enough to ask “What the fuck happened to Bill?” Telling them in advance never worked. They would just be drunk later.

It has been said that Ukraine has the most beautiful women in the world and, for me, it was certainly true. That said, I was partial, of course, because my woman was Ukrainian. She was unlike many American women that drove me insane with their prima donna bullshit. These women often acted like cats. Maybe that’s why it’s called pussy, in that they wanted you to lay the world at their feet because they were great enough to get up by noon. They still weren’t in makeup yet, but hey, they were breathing!

The other side of dating outside your jail cell, or country, however you see it, is that you don’t spend a lot of time there. So, I was very seldom home in the USA. Such was the life of an expat and engaging in a relationship with a foreign woman. However, isn't it funny how many stories start with a woman?

So, the stage was set, my ticket purchased, my hotel booked, and my beautiful girl very happy. I was batting 1000. It was a rare moment and time to have another beer. After all, what else are you going to do with the beer you purchased and put in your fridge?

I was to arrive in Odesa, Ukraine and stay for about 20 days keeping in mind my previously mentioned 20-30 day disappearance rule. I had work to keep me busy while I was there and, more importantly, keep the money coming in. However, strange things were already afoot.

In August, the prior month, I had received a call from a headhunter. This was typical of a contract consultant, but this call was better than most because the man could actually speak English. Usually, the calls came from Indians who told you it was very important they talked to you right then, knew nothing of your bio, and tried to get you to fill out their forms for them. These, I just hung up on. However, as a consultant, one rule was you always wanted to look for your next meal before you were hungry. I spoke with the headhunter.

He had a PMO (Project Management Office) Manager position in Saudi Arabia for the Ministry of Interior. Like everyone else that called, he was in a hurry and needed to have someone there in about a week.

Having not heard anything positive back from him, which was typical of those calls, I decided to book my trip to Odesa. That said, I maintained a relationship with him and occasionally contacted him and the company he represented. After all, you never knew when something would happen, and it didn't hurt to have multiple streams of income (consultant rule number 2).

It has been said that when you chase too many rabbits you catch none. However, who wants to read about the fucking insanity of the obvious? This is a book about a life derailed by a war that cost many lives. This is about adventure travel. This is about the joy of Ukraine and Ukrainian women. This is a book about the chaos that is Bill. For those of you who knew of parts of my life and always said you wanted to read the book, well, I'll just have to put out a box set down the road. Then you can sit down with your favorite cold beverage and read about all of the things that have led to my damage.

So, as it happened in rare form, many things did happen at once. It changed my business direction once again, which, according to my horoscope back then, would be the last time, but the star reader could have been drinking also. This book is about the bedlam of that adventure. Frankly, I started writing this as it was unfolding.

So, if I make it back to somewhere safe, then the book will be finished and all can read it. This was, of course, pretty much in real time through the love of the internet and beautiful places like Amazon. I don't mean the motherfucking jungle or something. I mean the website. If you don't see the end of the book, then like many bad movies, someone just needs to take what I have written here and finish the damn thing because I am dead.

I will, however, most likely be finishing this book though. It is in my nature. In fact, my vice president and good friend shared an email thought of me that is relevant here. It went something like this. "I don't know how you have managed to live this long with the choices you have made, but if you make it out of this one alive, I am going to nickname you the cockroach." Cockroaches are, of course, supposed to be the only things that are capable of surviving a nuclear war. Maybe it's cockroaches and Bill(s). We will just have to withhold judgment until judgment day. I sure know I have withheld judgment a few times or at least good judgment.

That being said, before we jump into all the things that actually make this book a book, let us first focus ever so briefly on this concept of adventure travel. I knew many people that would have considered adventure travel to be showing up at mom's house on her birthday hammered and others where adventure travel was being able to make it to the golf course and dodge little old ladies in golf carts. These people need not apply.

Others saw adventure travel as flying into Venezuela, BASE jumping off Angel Falls, and river boating back to Caracas all within a week.

These guys were much closer, but still had adult supervision and plenty of medical coverage if, for some reason, their guide could not keep them from fucking themselves up. At the very minimum though, like a drill instructor, he did give them someone common to blame, even though they booked the trip because they were having a mid-life crisis.

No, the adventure travel in this book is, indeed, pack your bag, go it alone, "Hey mom look; no hands," kind of travel that many will never know. If you get in trouble, you have yourself. If you need to borrow money, you have yourself. If something bites you in the desert or you get a cold, you have yourself.

It was the kind of travel where people resigned their positions on an almost weekly basis. Quite frankly not for pussies, but, let us hope that it is all good book material. That way, you gain pleasure and I gain money. A shameful plug of sorts I know, but certainly setting the tone of the book.

Adventure, in this case, is having the balls to go where life is pushing you, stepping up, going there, and then trying to find some way to enjoy it when you are there regardless of the environment that you end up in. Mind you, all these situations may not have been right for the arena at the time, but what the hell are books for anyway? Like Patrick Swayze in Point Break, "It is that place where you lose yourself and you find yourself all at the same time."

In that sense, if you are looking for any kind of adventure in travel, you are obviously seeking something. Maybe it is something that takes you from loved ones and jobs. Maybe it is something that makes you ask if you chose the right career path, but at that very moment in time you became a seeker. You became the true embodiment of someone that can pierce their own boundaries.

With the stage set, we embark on this story that I share with you that covers multiple parts of the world and about two years of my time, not to mention a number of people. I hope I describe it well and you enjoy it all, because it was a labor to live and a labor to write. I hope my work paid off for your education and enjoyment. Let’s get started.

Chapter 1: Growing Closer

In the fall of 2013, I boarded a plane very early in the morning headed for Odesa, Ukraine. A long trip at that being it was 60 fucking hours. Hell, I had made it to Kuala Lumpur in 40! However, with 9 plus hours in JFK and more than 16 hours in Istanbul, it all added up.

You always know you are about to have a long trip when you get to the ticket counter and they see Odesa and ask, "Which flight are you taking through Dallas?" I was like, "Look, lady, if I end up in Odessa, TX instead of Odesa, Ukraine, we are going to have problems." She quickly saw the error of her ways and actually looked at my itinerary. I was on my way.

JFK was its typical, uneventful destination where I sat waiting for my plane and drinking beer at New York prices. Of course, I was mobile, but had to continue to look stationary, so I responded to email and took a few calls before boarding the plane for the trip to Istanbul that took about 10 hours. Why, in a major airport like JFK, you had to fight for like four fucking outlets to jack your computer into was beyond me. This was, of course, supposed to be the modern era and airports were keeping up, right?

INSANE IN UKRAINE

By the time I reached Istanbul, it was altogether a different adventure. I didn't have a hotel there, Turkey required a visa, and since I was staying so long, I needed to collect my luggage, which of course meant clearing customs. So, in typical international money grab fashion, I paid the $20 for a visa so that I could get my bags and get a room at the airport hotel. After wandering the damn airport for about an hour, I finally found out where the hotel was and got a taxi.

It should have been a limo. As many of us know, airport hotels are not that far from the airport. In fact, often they are actually connected to the airport to make it easier for travelers. After all, airports are about travel, right? Now, when they were not connected to the airport there was still usually a shuttle or something to get you to the hotel. The fact that I even had to take a taxi was kind of stupid. However, the worst part was the taxi charged me like $15 to go half a mile to the damn airport hotel!

It gets better. Generally speaking, the airport should be pretty secure, right? However, in order to enter the airport hotel, you first needed to do a bomb check of the taxi and then your body and luggage again. It definitely gave you that safe feeling, especially when an hour later 20 cars with red flashing lights were out in front of the hotel. I, of course, had the solution: return some calls, send some email, and raid the minibar. I had a flight to catch in the morning.

I did attempt to order dinner, but fell asleep before they arrived and never got up to answer the door. At least they didn't charge me for it, even though, I did still want that fucking steak sandwich. That is life for you.

After all of this trip time and re-checking my luggage for the flight to Odesa, I settled in for the longest leg of the trip; 1.5 hours. Arriving in Odesa put me at ease and made everything worthwhile. I cleared customs, gathered my gear, and walked out to find the beautiful smiling face that was always Natasha, at least until I pissed her off. At this moment in time, all the hassle of the trip was worth it. That sounds like I was in love and the truth is, I was.

Natasha was 40 years old at this time. She was a beautiful, smiling, blonde, size nothing woman that seemed to be always happy and playful. In this way, she was not much different than the retarded kid who didn't know he was retarded. They seem to all make your problems look unimportant because they can find the joy in anything.

As a former lawyer for the Ministry of Interior of Ukraine, Natasha was very smart, too. She knew some English and we generally spoke in a combination of English, Russian/Ukrainian, Spanish, German, and the occasional hand signals. Accents could still be interesting, but somehow it seemed good that you actually had to think before you spoke to your mate. However, we were, after all, a couple and still able to piss each other off. That's what make-up sex was for, right?

INSANE IN UKRAINE

It was another morning flight and while we did get a taxi and went to the hotel, we were not able to check in until noon, so sex was out for the time being. We took the bags to the hotel, left them, and went for breakfast. I was fairly certain that one of the reasons girls stayed so small in Ukraine was because you walked almost everywhere. It seemed like the whole place was a no elevator zone.

That meant dropping your bags off at the hotel required you to drag them up three fucking flights of stairs to check in and then another flight of stairs to your room. The last thing you wanted to do was go back downstairs to go to breakfast, but that was life in Odesa. After breakfast, you worked off the calories going back upstairs to your room and then finally sex. The place was not for the weak of heart, literally.

Natasha and I spent many days talking, staying in bed, and hitting the local sights and clubs. It was September, so it was still a gentle fall in Odesa, and walking down to the Opera House and Black Sea port area was enjoyable with many people outside. That time of year the streets were still filled with performers, outdoor weddings, and many tourists. Outdoor cafés were still enjoyable also. During the course of all our conversations, Natasha and I agreed that she would go to the USA for a while and stay with me at my lake house in Texas as we entertained taking our relationship much further. Little did we know at the time, but this decision started us down a path to a whole new adventure.

INSANE IN UKRAINE

As a Ukrainian, it was not very easy for Natasha to travel to very many places in the world. Unlike Americans, she needed a visa for basically everywhere. Many places wouldn't even give her one, but were very glad to take the application fee. The United States was one such place that required her to have a visa. So, we began our visa process like any normal, stupid couple; on the internet.

We quickly found the embassy website and the visa process online and printed out the required forms to complete. The process for us meant filling out the visa application, which was not necessarily easy, booking an embassy appointment in Kiev, going to a specific bank in Odesa to pay the $160 application fee, and getting two train tickets to go from Odesa to Kiev nine hours away. That sounds easy enough, right? As a party bonus, we got to have train sex and see the countryside! Train sex is much like a waterbed. Just jump on top and ride the wave in.

The day came when we were to board the train to go to Kiev and Natasha wanted to go to the grocery store (called magazine in Russian) before getting on the train. In typical American fashion, I asked, "What the hell for? We can just go to the bar car;" stupid me. There was no such thing as a bar car in these beer cans. Often, you ended up with four people bunked up in your "cabin" and the beds made an army cot look like a queen size mattress. Lucky for us, we were the only two in our cabin, so there was still train sex with our champagne!

We really didn't know where the embassy was, but we wanted a hotel nearby since we were arriving the day before. You couldn't take luggage or anything else into the embassy, so I decided we would find a Wi-Fi spot when we got there and book a hotel upon arrival. We went to a McDonald's with internet and booked a hotel.

Now, you need to understand that McDonald's in Kiev was like some kind of five-star restaurant. This place was three levels and packed. Hell, I am an American who has traveled the world, so I have eaten at a few McDonald's, but this place was off the chain. Something you need to experience if you ever get the chance. You even get the same stateside customer service, which, of course, means they fucked up my order.

Again, we arrived in the morning. At the hotel, we checked in, dropped off our bags, and went to the bar where I could do some work, and Natasha and I could have a few drinks. We managed to get some sightseeing in, but there really wasn't too much within walking distance, so we stayed in the hotel most of the time.

While there, I picked up the local paper that had President Putin on the cover and was discussing the pressure that Russia was putting on Ukraine regarding a potential trade deal with the EU. It was October and I let that one register with me and even took a picture of the paper with my phone. One good thing that was in the paper though was that the Titanic exhibit was opening the very next day in Kiev. Another party bonus!

Early the next morning, we checked out of the hotel and went to the embassy. We arrived early because it was an important meeting and we really didn't know what to expect. Natasha stood in line to go in, but since we had luggage and that was not allowed in the embassy, I had to stay outside with the bags. It was October in Kiev, which is colder than Odesa, so I started walking, thinking I could find a restaurant or something nearby to wait at. Once again, stupid American. I ended up walking around outside for about an hour freezing my ass off.

Now, if you are one of the stupid individuals that think a United States Embassy is for Americans, let me educate you. Most of the workers in a US Embassy are citizens of the country where the embassy is located. They are there to keep people out of the USA. The US Embassy in Kiev was no different. Everyone that Natasha dealt with was Ukrainian and, in only 10 minutes, she was promptly denied a visa.

The reason for the denial, based on the letter she was given, was that they did not expect her to return to Ukraine. Now, mind you, she had family in Odesa, including her 11-year-old son, mother, and father. Her family had lived in the city center of Odesa for four generations, and she even had extended family in Odesa. So, obviously, she had no plans of going back, right? However, in the land of one of the most corrupt governments in the world, stood a man outside the embassy looking for people leaving with these letters vowing that he could help you for a small fee. After speaking with the man, prepared to pay his "fee," it became apparent there was nothing he could really do, so we left.

So, let's recap that. Visa application filled out; check. Embassy appointment made; check. $160 application fee paid to bank; check. Nine-hour train ride and hotel in Kiev for embassy appointment; check. Thoroughly fucked ten minutes inside the embassy; check. Time for another drink.

Needless to say, we were a bit depressed and decided to lick our wounds over a few drinks. We made a few phone calls to see if there was anything we could do in Kiev and came up with nothing. However, I did remember that there was a Titanic exhibit during that process.

I was a six-agency scuba instructor and I also taught wreck diving, so it was something I really wanted to see. We decided to go. It was a multi-language 3-hour guided tour that in Ukraine cost us about $15 a piece. If you get the chance to see it, I would highly recommend going. Hours later, a stop at the liquor store and a taxi to the train station and we were headed back to Odesa.

During the train ride back to Odesa, we spent many hours discussing our options for staying together. We decided to call an immigration attorney I had heard of in Austin, Texas. He, of course, wanted to know if we intended to get married, which caused us to have even more conversations. It is amazing what drives some people to marriage.

Sometimes it is love. Sometimes it is stupidity. Sometimes it is great sex. Sometimes it is a jacked-up government.

We decided that we wanted to get married. We informed the attorney and he told us it would be easier and cheaper to get married in Ukraine and then get citizenship in the USA than to get a fiancé visa, get married in the USA, and then get citizenship in the USA. It sounded easy enough for us and we were off in a new direction to get married.

Being an attorney, Natasha was immediately depressed with all the paperwork this process would take, but we wanted to get married and tried to remain optimistic. I couldn't help but think, though, that if paperwork depressed her, she had chosen the wrong fucking career. We started at the marriage office because that seemed like a reasonable thing to do.

Now, getting married in Ukraine and getting married in the United States were two very different things. In the United States, you could get a marriage license in a matter of days. The attitude there was basically "If you want to live a life of misery, who are we to stand in your way?" In Ukraine, when you were marrying a foreigner, the thought was more like "We are fucking miserable and you will be too. You are not escaping, bitch!"

The marriage office informed us in Russian that in order to get married we needed four items:

Passport – Translated into Ukrainian.

Visa (a.k.a. Passport Stamp) – Translated into Ukrainian even though it was a fucking Ukrainian stamp!

My divorce decree – Translated into Ukrainian.

A letter from the US Embassy that I was not married – Translated into Ukrainian.

We needed a bar.

We had about run out of time on my current trip and wanted to avoid the stress of our situation for a while, so we decided to just relax and take some time away from our problem and enjoy each other. We went back to the hotel with champagne and meats (kielbasa) and spent some time alone. We also decided to do some touristy things around Odesa because it was still fairly nice weather. Of course, as is natural, while we paused, we reflected on the situation and came up with some other options and tweaks to our plan to explore. It was obvious though, that in any event, I needed to return to the USA for some paperwork.

In doing our sightseeing, like any couple, we always took pictures. It was usually just the two of us, so we seldom had pictures together. Now, I don't know if you have any familiarity with Ukrainian women, but they want to protect their status as the most beautiful women in the world. They are experts at looking beautiful and posing oh so correctly. You could take a picture of one of them in the middle of a trash dump and never see the trash.

However, at the same time, that picture may have taken six shots to get. After every picture, they would run to the camera to make sure that it was either right or delete it promptly. You've got to love a beautiful woman.

All that said, we embarked on a whole other photo experience since we didn't have many photos together. No, it was not sexual so get your mind out of the fucking gutter! The lawyer in Austin had told us that eventually we would need to show the USA pictures of us together for her green card application. At that point, it was game on for the beautiful Natasha. We had mom, the son, Natasha, and I all over the city with camera in hand. She was a very smiley, happy person.

It was a very nice day. We had lunch together, with photos of course, and we covered a good portion of the Odesa city square. Odesa has had many cultures in the city over the years and the area was filled with statues and plaques of notable people. Additionally, there was a playground by her son's school that we spent time at that day. While we were hanging out, her son, Volodia, worked on his English with me. He liked my Texas accent.

Odesa was a beautiful place with a rich history spanning many years. While in that part of the world, I could almost feel how young America was as a country. You could also see the aftermath of WWII in many regions of Europe. There's a huge amount of beauty there, and I don't mean just the women, but there were many places that were re-built from war damage.

Taking any tour bus and listening to history would describe to you the turmoil of the area over the years. An interesting aspect of Odesa history that many tours mentioned was that many of the streets had changed names several times as different cultures were driven from the city. It had been a city of political tension more than anything else over its history. Maybe this had something to do with its Black Sea port.

The port was a central area of activity. Ships came and went on a regular basis and shipping operations were a big piece of Ukrainian business, especially given the magnitude of the oil and gas industry in the region. Many sidewalk shops and street performers would set up by the port to make their money. Some men showed up with giant Eagles that people could have their picture taken with for a fee, and it was not uncommon to see sailors walking the streets on shore leave.

The beautiful women and low wages gave rise to sex tourism in Ukraine and certainly in Odesa. The average Ukrainian made about $400 a month US. Prostitution was illegal, but widely accepted. Hell, prostitution was illegal in Thailand also and look at that country. Many girls would have liked to find a good man and were very open to marriage with a foreigner, but there were, of course, many girls that just wanted to take advantage of men. In that respect, it was the same all over the world.

While girls would have liked to find a man for marriage, many of them would have preferred to stay in Ukraine and were not looking to leave. It was their home. They were aware that leaving was very challenging and knowledge of English was very important. There were very many strip clubs and massage parlors in the city. You could even call someone up to come to your hotel or flat. With girls charging anywhere from $50 for an hour and a half to escort agencies charging $150 per hour, there were always men in the city seeking entertainment.

Additional challenges for women in the area in finding marriage were that men wanted young girls without children and some knowledge of English. Many companies arranged tours for men to go from city to city and event to event to find women. This was another business for the night clubs of the region. Age was not really adhered to in clubs as it was in the United States and it was not uncommon to find underage girls in the clubs. Sex with underage girls was also against the law, but still occurred.

In this world, Natasha was very lucky. Having attended law school and succeeding in business, she was considered rich there. She hadn't worked for about six years and really didn't need to except for the occasional side job for extra money. The pressure of earning a living in a

Western country was completely foreign to her. She spent her days in the fitness club and with her son.

Hell, in that country, maternity leave was three years off with pay and you could take an additional three years without pay. Doctors were equivalent to most in Western countries and were free or would, at least, provide services for a small "fee." There were many reasons why women wanted to stay in Ukraine, and, as an expat, I constantly entertained it. Natasha would have certainly preferred we had stayed, although we did want her to be able to travel more freely than she could as a typical Ukrainian.

As I write this, America is still suffering hard times and many, including myself, believe that those times will continue and the country will eventually collapse in debt as we have seen across the world many times. In a global economy, you can expect a leveling of the standard of living, and when you are at the top, there's nowhere to go but down. Even there in Ukraine with low wages, 2008 hit the country hard. We are all tied together in success or failure now and must learn to work accordingly.

Natasha and I spent many days walking the city center of Odesa and preparing to navigate the obstacles for our future. We had both been married before and, in Ukraine, marriage and divorce stamps were actually placed in your "passport." This was your Ukrainian documents, not a passport for travel. While her divorce had been final for more than

10 years, she never got the divorce stamp in her passport, so this was an item that she needed to complete. Being the foreigner in the picture, my list was different.

I needed to return home and obtain an apostilled copy of my divorce decree. We did go back to the marriage office to ask if I could send the documents to Natasha so she could take them to the marriage office. That was promptly shot down. We were told that we both needed to be present. There went another attempt to try to speed things up. What was I thinking?

Additionally, at that time, we didn't really understand what the letter of non-impediment to marriage meant to us and why I needed it. I mean, fuck, a divorce decree was not enough? More research was required and this was certainly a potential problem if not completed properly.

We had already traveled to Kiev and now I was headed back home to try to collect the documents of the world for us to be together. It was apparent that things were going to be interesting. In fact, if this was the process to get married in the United States, then divorce rates would surely decline because only the seriously dedicated and fucked up couples would ever get married.

The day came for me to fly back to Texas and we went to the airport. You never really experience that trapped by citizenship and lack of visa feeling until you are both at the airport and only one of you can leave the country. I checked my bags, kissed Natasha, and entered the security line. Natasha and I watched each other until I was completely out of sight, and I proceeded through passport control. While I was disappointed that she was not going to be sitting beside me on the plane, I was happy that we had been able to see each other, grow closer, and enjoy our time together while looking toward the future.

Chapter 2: Back to Texas

I boarded the plane back to Texas going back the same way that I had come; through Istanbul, Turkey. Once again, I got a visa and went to the airport hotel to spend the night. Only this time, after my vacation with Natasha, I was much more tired than the last time. The trip back was actually longer than the trip going to Odesa that took 60 hours. Hard to believe, I know, but it happened.

Since I was so tired in Turkey, after raiding the minibar and sleeping, I awoke late. I quickly packed my shit and hauled ass to the airport, through customs, and to the gate where the plane was still sitting. However, I was informed that the gate closes ten minutes before departure, so I was thereby fucked. That said, never fear, "Sir, we have another plane right behind this one in two hours that we can get you on. Just go to the transfer desk."

So, I went to the transfer desk, gave them my best line of bullshit, and they put me on the next plane. No charge. Cool, disaster averted. However, what about my connection in JFK? "Well, sir that is a different airline. We can't help you there. You will have to deal with that one at JFK." Well, ok, fuck it. At least I am back in the USA. Let's just hope my bullshit works on New Yorkers also.

Well, of course, in order to board the new flight from Istanbul to New York, I had to go back out through customs. At least this time I had a current visa still and didn't have to buy another one. So, I went back out through customs, made the loop, and then returned through customs to get out of the country again and board my flight to New York. In this day and age of electronic passports, I couldn't help but think someone was sitting at a computer terminal somewhere thinking, "This is the fastest drug smuggler I have ever seen. Stop him at JFK."

Finally, through customs with a new ticket in hand, peace of mind of no added expense, at the moment, and a few hours before the next plane, it was time to find a bar and get some work done. Now, while the hotel internet was nice in Istanbul, I don't know if you have ever tried to get the internet at the airport. Let me tell you, it sucked. I fought with it for an hour and finally gave up.

Additionally, Istanbul, at least at the airport, was very expensive. The good part was it was one of the few places you could actually make one purchase with multiple currencies, which keeps you from having to exchange it. They didn't take Saudi Arabian Riyal (SAR) and they didn't take Ukrainian Hryvnia (pronounced grivna), but it was a good opportunity to get rid of Euros and any of those pesky US dollars I had laying around.

After some time and a few beers, I boarded the flight bound for JFK and my next travel hurdle. Arriving in JFK, I saw that my connecting flight had indeed departed. I made my way to the counter to try my next round of bullshit. This time I was not as lucky. I got the big "fuck you" and, even worse, was told that there was only one flight a day going from JFK to Austin. So, not only did I have to pay for another flight, I had to get a room and stay the night in New York, too. I informed the person who was to pick me up at the airport in Austin, found a room, got back online to do some more work, and drowned my gripes in Coronas.

The next day was pretty uneventful as I checked my bags at the airport and proceeded to Austin. Everything arrived well, including me. It had been almost 72 hours since leaving Odesa by the time I got home. Home was my house on a lake outside of Austin, Texas, which I loved very much, and it was located, like many, in a golf course community. Since it was a three-bedroom house, not counting my office I worked out of, I sometimes rented out rooms. Renting two rooms basically paid for the house and it didn't really hurt having people around to talk to. At this time though, the house was empty except for me.

I grabbed a beer from the refrigerator, checked on the limited grocery inventory, unpacked, and began to recover from the jet lag. The next day, I went over to a friend's house to see what I had missed while I was out of town. Bryan was 50 years old and an interesting sort. He was always chasing strippers half his age and in legal trouble one way or another. A boat broker by trade, he basically did anything to make a buck and avoid working. He also rented rooms in his house.

His house was very different than mine though. My house was filled with paintings, Egyptian tapestries, and had a regular maid. His house had a stripper pole, a bar, a pool table, and many pictures of sexual conquest. It made for interesting times for his roommates, and they didn't seem to last long enough for me to get to know them, except for one. His name was Scott.

At this time, Scott had been renting from Bryan for nearly a year. An insurance agent who was in his 50's and still was an avid skateboarder, he was recovering from a divorce after about 26 years of marriage. I guess after that much time, a life change can be kind of wild. At that point in time, his life was dedicated to getting high, nailing girls, drinking scotch, and hanging on the lake. In that respect, he was not much different than Matthew McConaughey's character in Dazed and Confused, just older. Oddly enough, Matthew McConaughey lived in Austin also, so I guess we were all in the right place.

I am fairly sure it was the weekend at that time or, if not, we acted like it was. In fact, most of the time, those of us who lived on the lake acted as if it was the weekend every day. There was our group that was generally considered the "bad" element in our community and then there was the much older and uptight crowd that was just there for a peaceful retirement. I think, at last count, there were only about three hundred people that lived there year-round and about one thousand lots with eight hundred homes. Many people just used their homes as vacation properties or just rented them out. We were the true diehard lake rats, who knew the summer and winter rules, which were very different, because we lived there all the time.

That said, I sat with Scott for a while having some drinks and catching up. I told him about the trip, Natasha, and listened to the latest gossip I missed while I was gone. It was the typical who-is-fucking-who and who-went-to-jail-lately-for-what type of conversation that had become almost normal.

Scott and I were both avid movie buffs, so I went through his stash to see if he had anything new that I hadn't seen. Between the two of us, we probably had two thousand movies. After a while, I went back home to go through the mail and get some work done.

Working for yourself could be liberating, rewarding, and scary as hell. I didn't know anyone that had ever done it without almost losing everything at one time or another. You were always either a couple of good moves away from multi-millionaire status or complete bankruptcy. After a while, I guess you just got used to it and didn't worry so much. It was a lifestyle choice.

You may have found out you couldn't hold a regular job or you just didn't want one, making it worth risking everything to do it your way with your company. Just about everyone in my family had had their own businesses, so I had grown up with it. It still took a lot for me to get used to and I probably never did get completely content with it. However, what real job could I have had where I could continue to work while spending nearly a month in Ukraine with my girl? Many of us in that lakeside community had our own businesses to have the lifestyle we wanted. We were always working on being business owners instead of self-employed; work less and make more, right? That said, it was not uncommon to work more on your own business than you would have at a normal job. You just got to work the odd hours that you wanted to while drinking.

I went through the mail in search of bills and angry people. By that time, it was wake-up time in Ukraine for my lovely Natasha. We had our daily Skype call and I returned to work. Much of my consulting was by phone or Skype, so I could basically work remotely from anywhere. That made life easier, because you never knew where the next gig was going to be or when it was coming. The training was kind of the same way, especially since we didn't promote our own events.

So, my team and I set out to become experts at internet video production and marketing to smooth out the revenue curve and make our business more predictable. Radical business changes are never easy, but I had a huge technology background to leverage and it was clear that this was where the world was going. Travel for training was almost half the training expense and time, so much more was happening online. This was perfect for an international company and we embraced it happily.

Our company never did anything half-assed, even though it may have looked like it a few times, so we embarked on a serious mission. We became WordPress experts and procured our own development tools, hosted our own websites, embraced industry leaders in internet marketing that just so happened to be living in Austin, Texas, and we hired one of the largest industry leaders as a consultant. We were very lucky that the people around the lakes of the Austin area, as well as in Austin itself, embraced the internet for business. One of the largest interactive events in the world took place there every year. It truly was a lifestyle city and a great place for business. I didn't move there for that reason, but embraced the community.

Things continued on in good fashion, and I fell back into a routine at the lake again. Natasha and I continued our research into the document that would prove I wasn't married anywhere else. We both thought that we had identified it, so Natasha made another trip to the marriage office to confirm if it was indeed the right document. We didn't want me to fly back to Ukraine only for another surprise. As luck would have it, the marriage office didn't have a fucking clue about what the document really was either and sent Natasha to yet another office.

All of the back and forth occupied a few more weeks of our time and we never really got a firm answer. We had to do something very dangerous. We needed to trust the US Embassy's answer. All of this continued to take place while I tried to locate my divorce decree. My copy wasn't going to work because it wouldn't be a fresh notary. To add to the problem, my bitch of an ex-wife and I lived in separate counties. In typical ex-wife fashion, she had to make it a pain in the ass.

The divorce decree, of course, gets filed with the county. We lived in Travis County when we were married, we received our marriage license from Travis County, and we got divorced in Travis County. So, the divorce decree would have been filed in Travis County, right? Fuck no! That's too logical. You have to think like an ex-wife. She was living in Williamson County when we got divorced and her attorney was from Williamson County. Why the fuck should she inconvenience herself by filing in Travis County? She had filed it in Williamson County.

This is how time passed for a while; quiet, and pretty much uneventful. Of course, that can't last, right? As chance would have it, like every weekend, Bryan was throwing a party over at his house.

It was not uncommon for a party to go all weekend. It was also not uncommon for us to get knee-crawling, commode-hugging drunk during these parties, so driving wasn't an option. It was either pass out or walk home. Well, hopefully, home.

It was always possible that one may pass out in the yard, or the golf course, or even take a wrong turn and end up in the lake. When I say yard here, I don't mean the yard where you lived or even where the party was at. It could have been anywhere.

We often got together for dinner and drinks and just generally hung out until everyone fell over. The next day we usually made mimosas to recover before pretending to be respectable business people on Monday. This particular time, there were probably about six of us, counting myself. It included Bryan, Scott, Dave, I, and various others I didn't really know.

Dave was in his 40's, had his own home automation business, and a business partner. He didn't live on the lake, but would, from time to time, stay at his parents' vacation home in our community. By the morning of the next day of this particular party, Dave and I were tired, but still holding on after everyone else had gone to bed. We were just sitting on the couch talking and that's when it happened. I stood up to do something, I can't even remember what now, and, in a very uneventful fashion, I got my feet caught up underneath me and fell onto the carpeted floor. I hopped right back up and went on my merry way. No big deal, or so I thought.

While sitting back down on the couch, I noticed that my knee was beginning to swell. I am not talking about mild swelling here. I am talking about super-human, this can't be possible, grape-fruit size swelling. Dave and I kind of looked at each other in wonderment and I decided to do what any drunk man would do with a drink in his hand; put ice on it and cowboy up. That didn't have the desired effect, and my knee kept swelling.

Dave and I decided that we needed some adult supervision after that, so we woke up Scott. Scott came out and quickly said, "Ah, hell yes, you need medical attention." So, Dave and Scott loaded me up in the car and we were off to the nearest doc-in-a-box. One of the other usual side effects of owning your own business was that you didn't have health insurance and Obamacare sucked.

We arrived at the doc-in-a-box, but they wouldn't take me because I didn't have any identification. Like, why would I have fucking I.D. just to go to my neighbor's house? On top of that, they said my knee was going to need X-rays and they couldn't do them. What kind of medical facility couldn't do X-rays? Hell, every doctor's office could do those. So, we were off to the full-service medical establishment.

Upon arriving at the hospital, still obviously drunk, they began to take X-rays and asking the usual questions like "Were there any drugs or alcohol involved?" We, of course, gave them our best dumb look and said "No." The X-rays showed that my kneecap was not busted and was in place, but I was obviously in some pain at this point. They immobilized my knee, gave me a tetanus booster and a shot of morphine, and sent us down the road with a script for pain pills. There was the usual conversation that always falls on deaf ears where they told me to go see another doctor because I may still need surgery.

So, Scott and Dave loaded me back in the car where the night and the morphine finally caught up with me and I went to sleep after puking. As typical good friends, they stopped for more beer and to get my prescription filled so that they could steal some of my pain pills. After that, it was back to Bryan's house to be laid up on the couch. Now, I don't know if you are aware, but most lake houses are built to take in the view, so in a hilly area that meant they customarily had stairs. That basically meant I was screwed for a little while because I couldn't get up and down stairs or take care of myself.

We had a pair of crutches within the tribe, so I got around on those a bit and the party continued. I stayed at Bryan's for a day or so and the guys kept me fed until I could finally get around enough to go home and hobble up and down stairs. When I finally got home, I was ok getting around the house, but I still couldn't drive, especially a convertible Mustang low to the ground. Seeing my predicament, the guys and my assistant continued to bring me food until I reached the driving stage of disability. Luckily, this was only a week or so.

Of course, during the hospital episode and the time laid up at Bryan's, I was not online with my computer, but was still able to skype with my phone. Natasha figured something had happened and was concerned as to what. I explained to her what had happened and, like any smart woman, she became concerned that none of us had actually grown up. Natasha was very caring and even from thousands of miles away she just kept after me about the doctor, doctor, doctor. This never really stopped, but I guess I never really stopped getting hurt.

As is often the case with an injury, you usually end up hurting something else. I was not sure what I needed to do, but during the early stages of my knee injury, I was still having problems getting in and out of cars. I had had many injuries over the years, some of which had never completely healed, and my right shoulder was on that list. So, as I was trying to pull myself into a car, I managed to pull out my shoulder. This turned out to not be a major problem but still was painful for a few months and made getting around with my knee even more interesting.

The healing process continued fairly well, at least from a mobility standpoint, so I continued to blow off seeing the doctor. When the bruise finally settled in, my whole leg was black, including the top of my foot. You'd have thought somebody ran over it with a car. However, given the progress I was making, I figured it was going to be the typical "Don't use that thing for 6 to 8 weeks" type of deal.

I continued to work and evaluated Halloween costume options given my new leg and the fact that it was October. In the end, I did what I usually did for Halloween. I turned off the lights in the house, avoided kids at all costs, and drank beer.

Austin had been able to attract many major events like music festivals, tech conferences, and even the X Games. However, the one that I had been missing and really wanted to see was the Grand Prix that was run in November. Unfortunately, given my condition, I was worried about missing it again.

As the Grand Prix date grew closer, I became more depressed. It was obvious that if I were to do the responsible thing considering my knee, I would have to sit it out again. Yes, I was sometimes responsible. It came from being an international project manager and business consultant. This book just happens to be about some of the weirdness that has gone on in my life.

Upon the date of the Grand Prix, my friends tried to come through for me again knowing how much I wanted to go. They offered to get a wheelchair and actually push me around. Now, the Grand Prix grounds were laid out in a multi-tier fashion. So, my first thought was, while a nice gesture, that would be a pain in the ass and, I probably wouldn't be able to see anything anyway. My next thought was probably more realistic as I considered my drunken friends would be wheeling me around this multi-tiered area with speeding cars going by. I decided to pass.

In typical sour grape fashion, I stayed home with some beer and didn't even watch it. Instead, I did some more work. Working from the house, as I did, could be pretty lonely and closed up, which is one of the reasons we tended to mingle and party so much. However, occasionally, I would go down to the golf course that was within walking distance, get a bottle of wine, use their internet, and talk to people a bit. Hell, I barely used it for anything other than that, although I did play golf from time to time. This was one of those days where I just worked, talked, and drank.

I took in some sunshine, skyped with Natasha, and spoke with some people I hadn't seen in a while. They, of course, wanted to hear the stories of woe involving how I tore up my knee, but I had nothing cool to say since it was so uneventful. I wanted to at least say I tore it up at a naked hot-tub party or something, but no such luck. I had to break down and go with the truth.

The Grand Prix passed without me shedding too many tears and I continued to get movement back in my leg. Since it was November, Thanksgiving was upon us and I didn't often spend it with family. Natasha was in Ukraine and my only family in Texas was my father who lived too damn far away for my leg to be making the drive. So, in what had almost become traditional fashion, Scott and Bryan decided to cook up Thanksgiving dinner for the rest of us degenerates.

They always cooked up a storm when they decided to cook. They couldn't hold schedule for shit, but that probably only bothered me as I was the project manager in the crew. I hobbled my ass over to their pad and we spent time drinking and eating and, of course, checking out football on TV. We did the best of everything and it was another multi-day event.

Dave was also back in our neighborhood and joined the party. Since we did live so close together, it was not uncommon for the parties to move from house to house, especially when you needed something that someone else had at their house. Also, because we were a golf course community, most of these short excursions were completed with golf carts. This meant that since I wasn't walking so well, I ended up in the back of a few golf carts.

As the holiday weekend rolled on, we ended up at many houses for many festivities and, of course, many drinks. One morning, I am not sure which, I awoke at Dave's house on the couch wet. I pondered how I could have gotten wet since my knee was in such poor condition. As a former scuba instructor, I had woken up wet many times, but never when I could barely walk because of injuries.

I had a couple of extra bruises, but nothing extreme, especially for me. I, of course, decided to quiz Dave. As is customary at a lake community, even in the winter, we ended up lakeside. Apparently, we had taken yet another golf cart and I decided to catch some sleep in the back of it on the ride to the lake. Upon arriving at the proper lakeside destination after a very bumpy trip, my friends found themselves to be one person short; me. The trip had managed to bounce my ass right off the golf cart and into the nearby water where comfortable, but cold, I continued to sleep until they doubled back and found my ass.

As we were approaching the end of our holiday weekend, Dave and I decided to go back to Bryan's house to see what food was left for nutritional purposes and find a beer or four to help prepare us for the upcoming week. We ate, watched more football, watched some movies, and completed our usual weekend wind down. By this time, my knee was starting to get much better and I was even walking without the brace. This turned out to be a good thing because the next phone call would lead to more unexpected events.

Chapter 3: Saudi Detour

When you lived my life, you pretty much had to act like a Marine on a rapid deployment base. That meant being prepared to go anywhere at any time. When it absolutely, positively had to be destroyed overnight, accept no substitute! Lucky for me, I was born on Hell Island. For those of you that don't know where that is, it is Parris Island, South Carolina, which was a Marine Corps recruit depot. When the call came in that day, I was already re-planning my escape.

The call was from the Saudi guys that I had kept in touch with since August about the Ministry of Interior gig over there. They had finally gotten themselves in enough trouble that Ahab the Arab approved of me to go to Saudi Arabia to help. Additionally, in typical fashion for professionals, poor planning on their part seemed to them to mean an emergency on my part. They wanted me there yesterday like I didn't have a life or my own company that they already knew about. However, as I said before, I was calculating the re-plan while on the call. It was a re-plan because since I hadn't heard anything from them, I had already booked a 3-month trip to Odesa, Ukraine with all my documents to get married.

As they explained their situation to me, I could see they were yet another group of people looking for a savior project manager that usually ends up being a scapegoat because they are already fucked. The upside was it added to my bio, got me across the pond, and regardless of how it went down, put more money in my pocket. Additionally, there was yet another business upside. My company wasn't a Saudi Arabian company and this project would give me a 5-year visa for Saudi Arabia that could be great for business. Having a 5-year visa, you were on a very short list that made you look very good for Saudi projects because you were already packed.

As they continued their story of woe, I learned that they ended up finding a PMO Manager that was approved and actually started work. He lasted about 45 days before packing his bags, boarding a plane, and emailing from the United States that he had quit. These kinds of projects were not for the weak. You had to be a true diehard expat; enter Bill.

I explained to them quickly that not only did I have a house that needed to be sustained after my departure, but I had already booked a flight to Ukraine for several months to get married and have a honeymoon in Greece. I even mentioned that I was already on a project that was wrapping up and could be handled remotely while managing their effort. They responded in typical savior project manager fashion. They bought their way out of it by offering me even more money than I asked for.

Their political play was that I needed to be there quickly to make it look like they had everything under control. That said, I told them I would only give them one week in the country before taking two weeks off to head to Ukraine as planned. Of course, I also said that this would be on their dime since I booked my flights online and that money would all be lost. Saudi Arabia may have been dangerous, but I thought a pissed-off Ukrainian bride-to-be would probably prove to be more dangerous once she was told I would be making a slight detour to Saudi. I wasn't going to take any chances.

With the business end of my emergency change dealt with, I now had to deal with logistics. I had a house, checks coming to the house for months to come, and many bags to pack. This wasn't going to be a one-week trip. I was leaving for what could be many years or forever.

Shutting down the house really didn't seem to be an option, so I decided to go another route. I needed a house sitter. I approached Scott at Bryan's house and told him if he paid all his own utilities he could stay in my house for free since I was going to be paying for it anyhow. With some back and forth and thinking, he agreed to house sit while I was gone. I mean, how could he pass up free rent and having the whole house except for one room that I used to store my stuff?

So, we were off and running. Scott was moving in, I was filling out forms for visas and everything else, and I had a storage room to fill up. Oh shit! New wife before I get home. I needed to update my will and power of attorney and get some more fucking beer! It was chaos as normal. After all was said and done, I had left for life, with no real plan to return, more quickly than it was going to take for me to leave for Ukraine originally. All in all, it took just over a week.

It is an interesting feeling when you completely love your home and feel that you are where you are meant to be and then things instantly change. In a matter of moments, your life is very much different than what it was. You don't know if you will actually see home again. However, you are ready for everything that comes next.

It has been said, most people only move like 50 miles from their home where they were raised. I barely even knew people like that. I guess you are born into a life that is yours and you either own it or fight it forever. After all, when you hit that plane on trips like the one I took, you are leaving home and friends and family and everything else behind. You have to count on yourself.

Count on myself, I did. I didn't even remember the route I was taking to get to Riyadh, Saudi Arabia at 2 a.m. After so many flights, I think everything just looked like an airport or hotel. My father had often said that I had probably spent more time on planes than most pilots. Too bad I wasn't on all of those planes when they landed, but that is another book.

Landing in Riyadh, I was sure some people wondered who I was. I was dressed in torn jeans, a t-shirt, lake sandals, an Aussie leather flop hat that I always wore, and carrying my bag of critical gear. I was met by two men that I had talked to on the phone, however, never seen. I must have lookcd like an impressive leader at that point in time.

The first man was John. He was about 50 years old, former special forces, large, and usually angry about something. John was essentially the country president for the company and the man I would need to report to. You could tell he was used to getting his own way by being angry and threatening, but I was more than used to this growing up military and being born on Hell Island, so I took it all in stride. I had seen this all my life.

The second man was Allan. He was my executive officer (XO) and right-hand man. He would be the next man in line after me. Allan was originally to take the position I was filling and had even covered the role on an interim basis prior to my arrival, but wasn't hired for it. He didn't seem to let that bother him much. However, he too, like many in Saudi Arabia, was a bit of a hot head and would get mad about shit easily.

Allan was a retired Army Bronze Star recipient with many years of experience in Saudi Arabia. He was in his 60's and definitely knew how to get around in the dirt with Arabs. He was overweight, but still played golf occasionally. He seemed to enjoy Saudi life and it was probably because it was familiar and paid well. The man was definitely old school.

He was always early to bed and early to rise with a cigar in the morning and evening and occasionally during the day when he needed to detune. Allan knew everyone that was worth knowing. Both he and John had wives back in the USA that they barely saw, but kept leaving them behind for the life. He was also a good man and friend.

We made our way to the car, loaded up, and headed for the compound. Security was always a concern for Americans in Saudi Arabia. While many lived within the city, Americans were often placed in Western compounds that had much more security, which included armed guards, bomb checkpoints, suicide barricades, and razor wire.

Despite the extra security, the compound had lost 3 guards or so a few years earlier. During that same time frame, a compound near to us was so destroyed they made a movie about it, which was released in 2007, called “The Kingdom” with Jamie Fox. The hardliner Muslims were always killing infidels and the movie was based on bombings in 1996 and 2003. I was responsible for our men at all times, including at the compound.

Saudi Arabia was definitely the “Kingdom” of Saudi Arabia and they made sure you knew it. Most Americans there just called it the Magic Kingdom. The rule was you showed up with 2 bags; one for cash and one for bullshit. When the bullshit bag outweighed the cash bag you left and, of course, most eventually went back.

I didn’t believe they went back for the money. It was just a sickness within the people that did it, including me. Having any experience in the

Kingdom and a pass to go put you on a short list. A typical interview question was "Are you fucking crazy?" The ones hired were the men I was paid to lead.

We arrived at the compound and unloaded. It was an early Sunday morning. In the Magic Kingdom, the work week was Sunday through Thursday for religious reasons. I was still jet-lagged, so I spent the rest of the night unpacking and getting settled in while the men told the Ministry I had arrived and would make an appearance on Monday. After settling in, I got some sleep and tried to adjust to the time zone. The next day I was going to have to deal with those related to or associated with the country's royal family.

I already knew that the project and team were in trouble. There was only one way to hit a situation like that, right or wrong. The team had already gone through 4 interim leaders in four months. I knew it was not going to be an easy task to get them to believe that things could get better. With a team of international players, you had to go in strong with a boot in everyone's ass so they thought something had changed. It took me about 12 working hours to root out one problem and have him pack his bags for the plane in an attempt to get the team focused. Sometimes, you have to shoot a hostage.

As planned though, I had made my appearance for the Ministry and began to try to get familiar with the environment. Badge after badge, form after form, Allan walked me through everything as a great XO should, having been there before me. Everything was a bit of a whirlwind though. I had arrived from the USA in record time and was only going to be in Saudi for four days before going back to Ukraine. Before, I'd have probably stayed there, but I knew my life needed a change and it started in Ukraine.

In the end, I arrived in Saudi Arabia content and in peace. Somehow, someway, I knew everything in my life was changing and I was comfortable with it. I expected to stay many years in Saudi, but I ultimately expected an end to a life that had driven me into the ground. I didn't know how it would all work out, but somehow, I knew it would be good for me.

Of course, things never really work out as we envision. I guess if they did, people like me wouldn't have written books like this. Readers want to know about the places they have never been and the people they will never be. They are good people with loved ones, but their lives are certainly different than the life I represent in this book.

Ultimately, I did my short week, set up my office, became familiar with the men, unpacked, and started to dig into the contract. I unpacked at the house, but needed to repack for Ukraine. At least they knew I was coming back because I had left shit behind. As mentioned before, you just never knew in some of these situations.

Now, when you get to Saudi, you have to get the lay of the land. After all, the country is dry. That means no alcohol. However, some compounds had bars and the embassy always had a bar and pork. When your compound didn't have a bar though, it was time to improvise. This basically meant, "Who the fuck is manning the still?"

Now, when I say dry, I mean dry. You couldn't even get mouthwash with alcohol in it, so everything was like prison hooch options. The first big option was basically grain alcohol that would knock you on your ass. In Saudi, we all called it Siddiqui.

Now, there were a number of things you could do with this shit. The first was you could burn the fucking house down; or, you could just drink it straight and work on going blind; or, like many of the people that were there, buy non-alcoholic beer or grape juice that you could get anywhere in Saudi and just add the alcohol to it. The truly gifted learned that you could bring back Jack Daniels grilling chips from America and soak them in the Siddiqui and make Jack Daniels. Others would just pour it in a coke.

The other option was homemade wine. Me, being a connoisseur, did both. Basically, I boosted some 5-gallon water jugs from the Ministry of Interior office, bought some grape juice, put some sugar and yeast in it, covered the opening of the water bottle with a balloon that had a hole in it, and in a little bit of time, I had prison wine. Never try to keep Americans or Irishmen from alcohol. There is always a way.

Now, keep in mind, this was dangerous as fuck. The penalty for drugs and alcohol in Saudi was death. They cut your head off. I did hear of a guy that got prison, but apparently, he was needed to process payments somewhere, so they would let him out every few weeks to do that and then lock him back up.

Generally speaking, though, the whole Muslim thing was complete hypocritical bullshit. I have taught them all over the world and personally watched them. In most hotels, in places like Dubai, they had their own bar that I was invited to since I was their trainer. This was so they could drink, smoke, and pick up Russian prostitutes without being seen by anyone. Don't believe all their crap.

I think if more people knew the truth, the world would be a much better place. There's only a small group of really hardcore Muslims. The rest are pretty laid back. Even the women in Saudi discovered internet dating and Skype and tried to use it to get out of the country before it was finally shut down.

So, that was basically the first week. Orientation week if you will. However, it was approaching American Christmas time as well as time to go see Natasha in Ukraine. I say American Christmas, because Russian Orthodox was different. Their Christmas was in January. New Year's was actually the bigger day for them.

INSANE IN UKRAINE

I got back to Odesa, kissed my girl, and went to my favorite hotel. The hotel was going to be ok for now, but soon we were going to need an apartment. At this time though, I still believed I would get Natasha out of the country more easily. Stupid me. I should have known nothing was easy in my life.

As I said, it was December and it was Ukraine, so it was damn cold. We went to our favorite sushi restaurant, even though Natasha would not eat sushi, and got caught up. She, of course, wanted to know about the job in Saudi. Natasha was different that way.

Many women would be like "Hey, my man is making a lot of money. I'm going shopping." However, Natasha having seen bad times and bad things wanted to know why I was getting the big check. I guess it was good because she basically didn't want me to get fucking killed.

Of course, during the rest of our time together for those 2 weeks, we spent time with her family, including her son Volodia (short for Vladimir like Bill for William). Additionally, like most women, she liked to go dancing. Ukraine was really weird about this, actually, and some discos, which are what they were still called, only opened up during the winter.

I didn't mind though. Since it was not tourist season, the girls drank for free and Natasha got to hang out with her friend while I had a beer. The first time I went in though, security saw a knife on me and asked me to check my weapons. They said I would get them back, so I had no problem dumping them. The only thing was I had about 5 knives on me. They never forgot. Every time I went to the club after that incident, I got patted down. I mean, "Fuckin' A," I had just left Saudi.

On our list of things to do in Ukraine, was to walk through the document process of getting married. This meant going to more offices. We went to the marriage office, which basically sent us to the marriage police, and then I heard how my divorce decree was notarized instead of apostilled, so it couldn't be accepted. I called a friend in America and asked her to pick up another copy, have it apostilled, and send it to Ukraine.

On top of all of this, we had to go to Kiev again and fill out some more forms. By this time, the war was really getting going and we had to manage to get through all of it in the streets. On the 9-hour train ride to Kiev, we shared a cabin with an old Russian Commando.

Now, since the country was in a transition period, they were still also fighting over what the language should be. All the people I dealt with spoke Russian, but there were areas that spoke Ukrainian as a form of independence. However, Natasha did not speak Ukrainian.

We had to go to the US Embassy first and handle some forms and then to a Ukrainian government building to file the forms. All of this was fun shit that required alcohol. No wonder the Saudi's seemed pissed off all the time. They had to deal with the government too, but they couldn't drink.

When we got to the Ukrainian government building, it got even more fun. The forms said to fill out in Ukrainian and, of course, even though Natasha had been a lawyer for the Ministry of Interior of Ukraine, she didn't know Ukrainian. I simply told her, "You have to do something because I don't know it." She filled out everything and it seemed to be a passing grade.

They needed some time to process their paperwork, so we went for lunch and drinks. During the course of lunch, Natasha needed a chopstick lesson. She had all sorts of fun that day, or at least, I did. We waited about as long as we thought we had to before going back for our paperwork, and finally, the train again. We hadn't, however, factored in the political war that had Russians and Ukrainians killing people over "Western" influence of Ukraine while Natasha and I were trying to get married.

When it came time to wave down a taxi, it was fucking useless. Nobody was stopping. Natasha finally got an old man to stop and explained our situation and offered him $3. He got us as close as he could in a car, but we were on foot through the fires and shit the rest of the way. We almost missed closing time, but managed to pull it off. After that, of course, we needed to get back through the bullshit and to the train.

We managed to get back through the snow, the fires, and the people rioting and actually found some taxis. However, fuck if Natasha wasn't actually haggling over the price to get the fuck out of there in the middle of a riot while I was freezing my balls off. She blew off one taxi for just a dollar! When he tried to come back and say ok, she basically said, "fuck off," and told me she didn't like him. REALLY??!! I liked him at that moment! We were surrounded by people wanting to kill Westerners and the fucking place was on fire!

I have no idea how the driver got us through the crowds and the city. However, finally, we got to the train station just in time to haul ass through the store, picking up champagne and food, and, actually made it onto the train for Odesa. We were able to get out ok, but for a while, they stopped running the train to Kiev. This was not going to be your normal wedding situation obviously.

So, having skirted all of that bullshit, we decided to detune a little bit. After all, some of these people were pissed off about Western influence in Ukraine, which was the reason for the war, and here sits a Ukrainian woman and an American man in the middle of it, trying to get married. Not to mention, we were both former government peeps. Not the best situation for living a long life. I needed to return to Saudi soon, but we didn't want to be stressed out the whole fucking time. It was time for a drink.

Keep in mind that, while all of this was going on, I was still handling emails and calls, and managing a new team in Saudi I had barely gotten to know before leaving. The project was behind schedule when it came to fieldwork and we needed to get on that quick. Most people, for obvious reasons, didn't really want to go to the deserts of Saudi for work. Many were complaining. It was just something else I had to deal with.

This was a situation that was extremely delicate, but in a foreign country, delicate is not always the answer. I mean, I was already sitting in one country that was breaking out into war and working in Saudi, which had never been known for being stable. Natasha was not going to like my approach to this, so obviously, I wasn't going to tell her.

I was not the kind of paper-shuffling leader that enjoyed sitting in an office. If I was, this book probably would never have been written. I damn sure wouldn't have ended up in Ukraine, Saudi, or the rest of the world. I had a man back in Saudi putting together a field deployment schedule. He was one of the men emailing me, telling me of all the complaints. This had to be dealt with promptly before my return. I simply told him and copied everyone else, to schedule me as the first one in the dirt and the last one out. Anyone who didn't want to go could resign.

Well, it wasn't really the Christmas I had been expecting, but then, nothing really ever turned out that way. I was going to have to deal with

Natasha when she found out about the fieldwork in Saudi, but fuck, she was in a war zone anyhow. We were going to be screwed either way. Some things just went down that way.

Natasha and I enjoyed our final days together for a while and spent a lot of time in bed between restaurants. We talked about the future and, hopefully, getting her to America. However, as you will learn, that was going to be difficult. Luckily, she was a person that liked to travel, but her son, as with any relationship, posed different complications.

Chapter 4: Still in Saudi

So, I made it back to Saudi. At that time, we really started to knuckle down. It was going to be dirt time and everybody had to get their game faces on. They were worried, but that was ok. It tended to keep people focused.

The job, on the surface, was pretty straightforward. It was inspection, acceptance, and reporting of multiple facilities all over the country that will never be disclosed in this book. We also had the typical daily and weekly report bullshit, but I basically just had to review, approve, and sign that shit.

We did have operational plans that had not been prepared yet, so we had to focus on those as well. This was January and, in the course of screwing around since August, they managed to fuck it all up pretty good. On top of that, the US company had no real international experience.

This company had presented themselves as having international project management experience, but John realized the company had little to no such experience. He, also, quickly told them that sending people to a 1-week course to live in an established and protected camp was not international experience. The men on this project were expected to go into a foreign country with no assistance, set up camp themselves, and complete the project with no security. This was a much different situation that required different kinds of people. After all, our official visas from the government said we were not allowed to work there.

The situation was actually even more complicated than that. We couldn't open bank accounts. The company couldn't even be paid in Saudi because it wasn't a Saudi company. Additionally, it wasn't a credit card kind of country.

After all those months there, I still had to explain to the guys in the states that if they needed us in the field, then we needed cash for approximately 20 men for expenses. This seemed to come as a complete shock to them because, so far, the team had been doing the work and basically sending bank notes back to America via DHL. These guys were out of their league.

It wasn't much different than dealing with Natasha. She thought she had traveled, but all her travel had basically been a bunch of Russian speaking people on a tour bus. The guide would just sit there pointing out things of interest, saying "Look at this." That just wasn't the same thing as entering a country and setting up operations.

Unfortunately, it doesn't take too much for some who have traveled to a foreign country for work, to feel they have gotten in way over their heads. This caused problems. These are the people who didn't even know what they didn't know, but figured they knew it all. There has never been a quicker way to death, in my opinion.

In short, it is the rare people that just pack their bags and say to themselves, "I am going to go to a hostile place I have never seen before without a guide and figure it the fuck out after I get there." This doesn't happen often, but it was a trait needed for the jobs we did. Hell, just the food and making alcohol will change you. Always make sure you know what kind of detail you are signing up for, because on stuff like this, there is you and the team and that is it.

Basically, once we got in the dirt that became the full focus. Now, you may have seen photos of Dubai and shit and thought that was what the Middle East looked like. It couldn't be further from the truth. This was one nasty ass place with trash and poverty everywhere. Saudi's were just lifting up their dresses and shitting wherever they decided to. It is an experience you never want to repeat. Trust me.

Most people know of all the money the royal family had gotten and the kind of shit they had bought. However, they didn't realize that the rest of the country got shit and many just tin-cupped it in the streets. The difference between the poor and the royal family was so disproportional it was not funny, but nobody ever talked of Saudi. I'm not saying I believe in socialism, but that country was fucked up. I am happy for all the places I have ever gone to. I just am not in a hurry to go back to some of them.

Once we got the first dirt gig out of the way, we got on a roll. The project docs started getting completed. The people who weren't going to make the cut went home. We were making progress.

The ones that really didn't want to work or go to the desert in the end, were the Muslims, who were supposed to make up 1/3 of the team per the government. On top of that, places like Mecca could only be completed by Muslims for religious reasons. This is the nature of some projects in foreign countries.

In fact, there was one day when the Muslim team members sat in the car and refused to get out. They informed me that I was supposed to have provided them with clothing and gear for the environment. I said, "Fuck you!" These asses went all the way out into the desert and pussed out. I explained, in very simple terms, they could do their job or get the hell off my team. One of my other guys, an American, was not as nice. I had to break that up.

The sense of entitlement that some of them had was ridiculous. They fucking lived in that area of the world, yet they acted like pussies in the sand and rain. I only had my hat, scarf (shemagh), and coat. They could get with the game or be gone. They finally got to work, but it was always a problem. One guy even wanted to bring his whole family and have me provide a suite for them. NOT!

The first place we set up field camp was a hotel in Dammam on the East Coast of Saudi. It was only a taxi ride away from Bahrain where ISIS was at the time, but they sold beer in Bahrain. The upside was that in the field I could dump the office gear and tie. I always preferred my Indiana Jones look anyhow and being out of the office. This quickly became clear to everyone.

Ultimately, I caught hell for putting myself at risk. I was told how a General needs to be willing to have his troops get hurt or even die. I knew this already. I explained, however, that I just didn't lead that way. The work that needed to be completed for success was going to happen in the field, and I was the kind of leader that would never ask someone to do what I was not willing to do myself. Sometimes you just had to show the team that.

Now when you get in the field, things tend to change. Even though we were away from home in the office, a government building puts things in a different context. When a team is in the field and together all the time, nerves tend to get rubbed a little raw, especially when you have Muslims and Americans together.

Basically speaking, the Saudi's just considered Americans the high paid workers that they needed, doing what the Saudi's wouldn't do. After all, it was a country where you could work for barely 15 years, retire, and leave your job to your son, if he was qualified. It was a straight-up fucking zoo by American standards.

So, eventually, we reached that boiling point on the team. I came downstairs to the lobby of the hotel we were occupying only to find Muslims and Americans going at it because they had finally become upset with each other. It's bad enough when we are all in close proximity for long periods at a time and share the same values as part of the same culture. It gets different in cultural circumstances. I broke it up, questioned my right-hand man in the field, and chewed some asses. It is always to be expected, though. We were all living together in the field away from home and base camp.

So, after a while, this stuff starts to burn my ass. I had a new 2-day plan. Basically, I ditched the fuckers. I mean, there's only so much whining and crying a person can take. I figured we had two days off and if these guys beat the shit out of each other, it was not my problem. It was after hours and off duty.

The new plan was to go into ISIS-occupied Bahrain solo with no backup and in nothing but a taxi. Good plan, huh? Well, at least they sold beer. I booked a room online, ditched the team, told nobody shit, including Natasha, and jumped a taxi. The next thing I knew, the phone was screaming because the General went missing. Even Natasha was asking "What the fuck?"

So, I finally decided to take everyone's calls. I told them exactly what I had done and why I did it. I proceeded to get my ass chewed by everyone. At least this time, however, I was able to drink while getting my ass chewed. The operation was worried about their General and Natasha was worried about her husband-to-be. There was never an ISIS moment. I tended to find that people you were prepared for were not a problem. So, I did it twice.

Eventually, I drug my ass back to Riyadh, but was only there about a day. One morning, I was getting calls about problems in the field. I didn't drive in Saudi, so, I grabbed one of my guys and said we got to go. It proved out like I said before. The job was in the field and that is where they could find me.

See, what you need to understand about these "savior" leadership roles is that you will walk into a fucked-up situation and ultimately get the ax because they have to blame someone even if it comes out right. It is a bit of an oxymoron with emphasis on the moron part if you ask me, but I can't change the way life is. I just did the job and charged accordingly.

It was becoming clear that my time in Saudi would not be as long as I had originally hoped for, but the wife-to-be was another problem anyhow. There was a British man there that was basically my enemy, but we talked. He had married a former USSR woman too and we both wanted to get our wives into the country. Saudi's response was simple, "Hhhmm, I understand your problem. How about we give you another wife here?" Welcome to the Middle East.

You have to understand that in the kind of life I was living, everything changes. You have to take in new cultures and be prepared for anything. I'm not just talking about security, but languages and environment and the whole meal deal. When you change countries, it starts all over again. This was a completely different lifestyle than many lived.

For example, I was bouncing between Saudi and Ukraine. Actually, I was doing it every night since I was talking to Ukraine on Skype. So, basically while my day was spent dealing with Arabs, I was learning about my wife-to-be in Ukraine. This can be an interesting situation, and that doesn't include dreaming in three languages.

If you are going to travel the world, you have to become versatile at dealing with different cultures. You would probably think that Natasha was good at this, but she was one of the worst. She knew her former USSR scene and that was about it. We will talk about Moldova later, but even when we got there, she thought it was a foreign land. Moldova was only about 30km away from Odesa.

So, we stayed in this mode in the field until the first security problem appeared. The military rolled up on us and extracted us from the area via escort rather quickly. There was shooting not far from our location, so they pulled us out to keep us from getting killed in the country. Not to mention, possibly ending up with a shit ton of bad press.

We generally worked closely with the military since we were the Ministry of Interior and on a security-based operation. Additionally, from time to time, the American Muslims would take off for some sort of holiday without telling anyone, so that shit had to stop. Hell, they didn't even know how to fight.

Now, I know what you are thinking since I already told you about my solo Bahrain trips that I never let anyone know about until they found me missing. However, as it has been said many times, rank has its privileges. When it came to cold beer, I was going to exercise all the privileges I had coming. Everyone else got grounded.

I mean, they could have free movement around the city and shit during downtime, but hopping a plane to some other spot would not be happening anymore. This was a difficult situation since they were basically traveling for religious tourism. However, they could listen or go home. After all their complaining, they complied.

The America Muslims were an odd sort anyhow. I mean, in general, Muslims could seem odd to begin with. This is true even if they are one of the good ones. However, the American Muslims didn't even want to follow their own cultural norms practiced in Saudi.

We once had a traditional meal with high ranking members of the military and descendants of the royal family. This consisted of rice and lamb, ditching your shoes, sitting on the floor, and only eating with your right hand because you use the left to wipe your ass.

It was custom for the men to eat first, then leave the room. The women would enter next and could have anything left. Then, it was finally the servants turn. Nobody intermixed. However, as soon as we sat down, one of my American Muslims was like "I need a fork." I told him to sit his happy ass back down. When you eat with one hand and need to pull something apart like lamb, then somebody else holds part of the food and you use your one hand to break it apart.

He survived his ordeal and we embarked on a series of meetings to try to figure out some problems. Saudi's were not the best at this. In fact, if you have seen a movie about business in Saudi, you may have learned that just because there is a meeting scheduled, it doesn't mean they will show up. They tended to take off at any time for any reason.

All along, we were dealing with people who resigned and needed to be replaced as well as others that had to be rotated out. Some had to rotate because of their visa requirements and some rotated to have some family time. Some visas were more difficult than others, but some people could literally leave the country and come back 10 minutes later. It was a challenge to keep up with it all.

Meanwhile, I still had shit going on in Ukraine and had to keep Natasha on task. My divorce decree finally arrived in the proper format, and Natasha went to make sure it would be approved. She continued to complain about doing the paperwork to which I questioned her choice to become a lawyer, again.

Alan, my right-hand man in Riyadh, rotated back to the states for a bit where he had two houses. He was beginning to have medical problems and we all had growing concerns, but he was a tough guy that wanted to stick it out. It was always good when we had someone rotate to the states so that they could play mule on the way back for others. I had ordered our wedding rings, but had to leave before they arrived. Luckily, Alan was able to bring them back for me.

Political issues were becoming worse, also. The company had kicked off the project in August, but they did it without a PMO Manager and an established budget. Some of the guys were actually funding large piles of cash and getting very tired of it because the company didn't know what they were doing. On top of that, the company was pissed off there wasn't one complete budget for what they spent in the US as well as Saudi, even though they had created the fucking problem. So, stateside people were being replaced too.

The two Brits continued to be our obvious enemies, but that was something we had to contend with. Meeting after meeting, they were in blocking mode. Delivering on actions was really not on the top of their list. In fact, before I got there, the team couldn't basically get them to do a fucking thing and we weren't even fully staffed up yet. The Saudi's were going to grow tired of the US operation and lack of support and planning. The project was over $1B.

These were the kind of things that had to be managed, though. Hell, I even needed to take care of room assignments in the compound. One guy managed to get his wife into the compound. Of course, I heard about that all the time. He never went to the field, so he was going home.

We lacked the country documents for things like computers and phones we didn't have, but needed. Alan was basically the only guy with the right docs, but he could only buy so many items. The other piece of bullshit was that we always needed to upgrade the drivers for the cars we had. This was a true cluster fuck.

Well, cluster fuck aside, we had work to do until it was done, or, I was gone. During this time, it was beginning to come out about Saudi's involvement in the 9/11 attack on America. That didn't help our status being there to ultimately protect them. Many in America frowned upon us.

Ultimately, that was a political issue and I wasn't much of a politician. I was staying at the compound for a while. We were trying to make it more like home, or at least as much as we could. This meant we were finally doing grill get-togethers and shit.

Chicken, steak, and homemade alcohol were always a great combo. The compound was pretty well-equipped. It had a bowling alley, pool, jewelry store, restaurant, and a small market. No bar though. Nobody really used the amenities. We just selected which house was going to be hosting and took our downtime.

We continued to battle it out over how the reports should look since the Saudi's kept changing their minds. It was driving everyone nuts. We were making progress, though, on other documents, so that was a good thing. Hell, they were even thinking of changing all the fucking technology as we were going through acceptance of the first batch of shit. They really couldn't stick to dick.

In Ukraine, things continued to heat up. It was unpredictable how it was going to go down. Kiev was still occupied and on fire and President Yanukovych was feeling the heat. He was basically the face of corruption in Ukraine at that time and everyone had seen enough of it.

Building a new country takes a long time, but it doesn't mean that people will be patient. They are more concerned with what they don't have. Natasha grew up with the violence, so she was pretty much used to it and didn't think about it much. She wasn't a fan of the president, though, and felt he didn't speak well at all. She would say, "He doesn't know Russian or Ukrainian. How did he get to be president?"

In that respect, I guess we have all had world leaders in our respective countries we didn't really like. Hell, even Saudi, with all of their oil money, was beginning to feel the financial pain of a slowing economy.

They were booting out foreigners to create jobs for Saudi's. In fact, it was becoming even more difficult for foreign families to come in. It used to be that the whole family could be in the country and the kids could go to international schools. All of that was clearly changing.

The day eventually came when I was to be replaced. I had basically made the average time for the last PMO Managers, even though it seemed short-lived. When these things happen, they are basically like "I got a ticket for you to go anywhere in the world that you want, so where am I sending you?" Of course, I said, Ukraine.

They basically asked the question about three times because they thought I was out of my fucking mind. Everyone all over the world had been watching Ukraine burn and get worse. It was not a place that Americans should be flying into solo, but it was something I had to do. I wasn't going to let a little civil war change the plans Natasha and I had to get married. Especially since she was the most beautiful thing I had ever seen in my life.

I had collected enough money to try to work something out in Ukraine for a while, but I had no real plan. Natasha and I were caught by surprise on the timing of the situation, so she was actually off on a train to a different city with her mother for a family event. I told her I was headed back and would be staying at our favorite hotel. Of course, I wasn't going to get any help carrying bags up the fucking stairs this time. The city needed elevators.

The growing conflict in Ukraine was beginning to attract foreigners from all over the world. Ukraine's military was not prepared for conflict, even from its own people. Also, many countries had an interest in how the country and any new government might turn out. During conflicts like the one that was growing, it can become a bit of a feeding frenzy.

There is always talk of foreign government involvement in things like this, and in a lot of ways, there always is, if not direct involvement. Ukraine has always been strategically located, which is why it has been ruled by so many other countries. Russia did not want Western influence in the country though, so me showing up alone could have been dicey and was well-known.

I had a few days left in Saudi before I could board a plane and get out, so I did as much of the normal housekeeping tasks as I could and drug ass packing. The rest of the time, I watched movies and got naps. I needed to be well-rested when I got to Ukraine. There was no way to know what to expect.

Eventually, I got packed and said goodbye and good luck to everyone. I was leaving basically the same way I came in. It was a red-eye flight. At least by the time I got to Turkey, it would be beer time.

I would be arriving in Odesa a few days before Natasha. Although I would be able to get a little more rest, I couldn't wait to see her again and start our lives together. Most people never know how things will turn out, so not having a plan was not viewed very much as a problem at that time. Once again, I was off on another adventure, but it was what I had been trained for.

Chapter 5: Back to Ukraine

Well, I arrived back in Ukraine in body shock. I mean, I had just left the desert, it was still February, and my cowboy boots were sliding across the ice. In the interest of the family, you would have thought that at least Natasha's sister could have kept me warm, but she was a quiet married girl.

I checked into my standard hotel at the time and, lucky for me, the minibar was already stocked. After Saudi, I was dragging ass. I didn't realize how much it took out of me or how tired I was getting of the life. Now, I had to start a new life in a new country with a house and bullshit still in another country.

However, the food was a fuck-load better in Ukraine. I wasn't eating camel ass anymore and, hopefully, never would again. I had acquired a couple of favorite restaurants, and one was run by Natasha's cousin, so at least people knew me. I banged out some food, rested, and waited for my girl.

You never really know how tired you are until you lay down after a gig and have a hard time getting up. You just take a pull off the beer sitting next to your bed and then you lay back down. Now, don't get me wrong. I was not fluent in Russian, so I was learning on the fly every day. Thus, this was part of the experience.

So, I spent a few lazy days eating sushi, going to my favorite microbrew place for mussels, and, of course, hitting the Irish pub. There were Americans in Ukraine, most of them were illegal, and never wanted to be found again. Ukraine was not a country known for great relations with America to the point of deporting people back. In fact, they couldn't even legally ask for your passport in the streets. Many of us tended to find each other and it gave us a feeling of home.

Leaving your own country for another country is a personal decision. Everyone does it for different reasons. Some did it for work. Some did it to avoid criminal problems. Some did it just because they didn't like home or they like to travel. Additionally, some did it because the Ukrainian women were just flat-ass beautiful.

I met all these types there and they were all good people. We didn't concentrate much on each other's respective pasts. There seemed to be an unspoken rule about questions you could ask or not ask. This was a new realm and we were Americans in a former USSR country, which we had all been taught to hate.

Natasha finally showed up and it was a reunion with all the proper activities. Her family, except for Volodia and her mother, still knew nothing of me. Our relationship was a secret. We both worked or had worked for foreign governments and we had been taught to hate and kill each other. Life, time, and love changed everything, whether it was politically correct or not. It was time for us to start navigating some new problems for us.

Things continued to heat up in Ukraine with the war. Ultimately, if I had been a smarter man, I would have realized how dangerous my situation was. Nobody knew exactly where I was and nobody would be notified if I went missing. Natasha's family didn't even know about me while I was trying to learn Russian. If it went south, I was just dead and gone and it was getting worse by the day. However, it was what I had gotten used to.

Natasha showed all the signs of being a great wife. She was happy I was home, worried about injuries I had acquired, and fought everybody over our money. She would fight a taxi driver over a dollar. She was very attentive, and like a typical Ukrainian woman, she always wanted to be the most beautiful thing on the planet. She never failed in these areas.

It was a different experience being in a foreign country, with a foreign language, and no plan to get home. This wasn't a tourist situation. There was no return ticket. There was no support. I was straight up ass-out-and-you-better-make-friends-or-you-are-dead.

I was reminded of this very clearly one day when I heard a big crowd outside the window. I had paid for someone to come over and get Natasha's nails all pretty. It was very important to her, and, of course, her looking beautiful never hurt my feelings either. I looked out the window to see what was going on and saw hundreds of people marching down the street screaming “Putin! Putin!” At this point, it really became obvious the situation was getting and going to get serious. This was going to be a war.

I still had my eye on the ball though. It was all about Natasha and our new family. Some people once gave me the nickname “Max” because I was either on (Maximum) or off. I have never stopped once I got on something. I had learned at a young age when you take failure as an option off the table, everything changes. It changed my life.

For me, this wasn’t an operation to come home from. I was home. It was my new home. Sure, I didn’t know the place or the language, but I could figure out who the good and bad people were. Additionally, many there wanted to learn how to speak English. So, between a little English and a little Russian, everything was fine.

When you are prepared for anything, then you can do anything. Most things in life turn out to be about confidence. That’s not to say you don’t still need paramilitary skills in a lot of circumstances to live, but it starts with balls and confidence. Natasha often looked at me like I had three heads. I could take a champagne cork to the eye butt-ass naked and then go out and beat someone’s ass if necessary. Yes, that shit happened.

So, it was kind of a “No shit, there I was” moment. We continued to stay focused on our mission to get married. Office after office and document after document, we marched forward. Again, it was a true cluster fuck. I didn't read Russian and Natasha hated documents.

At one point, we got our docs to the marriage office only to find out my middle initial was wrong in the translation. So, we had to go back to the translation office to get it fixed, but there was fucking snow everywhere. When we got there, I told them to fix it. "NOW!" Then, reprint it. I had paid for expedited service and was an asshole of a project manager that way.

They promptly told me that the person they needed couldn't walk to work that day because of the snow. In typical Irish fashion, I said, "Fuck you! Get her a taxi and get me the fucking doc I paid for in 60 minutes." Then I got a beer. Well, we got the taxi, but Natasha and I had to get in it to go pick the chic up and take her to work.

So, we got her butt to the office, she got the doc done, and we left with doc in hand. Now Natasha and I were back in the snow. We hauled ass across town to the marriage office and resubmitted our documents to the chica there. She finally said everything was ok and gave us a license and a date to get married. I paid $50, or whatever it was, and we went to another office that asked about photos and shit. We got the full meal deal.

Natasha and I had just completed a major milestone. A Ukrainian woman and an American man were now approved to be married in

Ukraine while war was breaking out. It was time to party some and we damn sure did. It had taken months just to get to this point. However, if it was easy, everyone would do it and there would be no reason for this book.

We grabbed her friend that did know about us and we went out. Natasha quickly learned that drinking with an Irishman is not something you want to do. She was still going to try to stick it out though. I gave her credit for that and, of course, she was sexy as hell.

Natasha spent a couple of days dragging ass afterward. We spent a lot of time in bed and planned our next move. Actually, I planned our next move. When Natasha was somewhat vertical, I got some champagne and food down her and started talking to her about the plan. I was going to need her Russian skills to make it happen.

Basically, the hotel was not a long-term solution. It was time to get settled into an apartment. Most places in the world asked for your docs (a.k.a. passport) to get a pad. Now was not a time to be flashing an American passport. Furthermore, I didn't even know how to find a place without getting ripped off. In short order, Natasha had us a maid/cook and an apartment. I knew she could do it.

It was about two blocks away from the hotel, but she liked being downtown. She was just that girl. She had a car once, but sold it to get through college. It really didn't matter. The woman had no sense of direction and would get lost in the city she grew up in. She only drove the car down one street, then turned around and went up another street. That was it. Her sister, Lena, was just flat crazy as a driver, but they were both still beautiful.

Sure, I tend to excuse everything with a sexy body and a beer, but I'm fucking Irish. I mean, I don't have kids of my own and I never will, so what the fuck else is there? There are only beautiful women, beer, and bullets. That's it, other than Molotov cocktails.

So, we got moved out of the hotel. I mentioned Lena above, but she still didn't know about us yet. It would be a few months before Natasha had the guts for that. We were in the danger zone from a Russian point of view.

Well, we settled in. Natasha would go see her friends, the protests continued, and I started my first book. We started to draw attention and I did some interviews for America. Many of my friends said that they were happy that I was telling them what was happening because American news didn't say shit. Some of them even happened to be Ukrainian Americans and damn good brothers. They always watched for me.

As time moved on, and Natasha and I got closer to our marriage date, she, of course, wanted her son, Volodia, to live with us. At this point, she kind of felt she had picked the wrong apartment. We worked out short-term living arrangements, but she started looking for another one. I was good with that because Volodia was a very sweet, innocent young boy.

He was a really great dancer, which in that part of the world, means acrobat. Unfortunately, times were changing. He went from wanting to dance to being concerned about his life. His dad would never really do anything outside with him, so the day finally came when he arrived home and asked if I would teach him to fight.

There's this strange feeling you get when a young, innocent boy finally asks you how to fight and wants to learn how to use a knife and shit. You don't want to be the cause of his loss of innocence, but you don't want him hurt either. I taught him. The streets were getting worse every day and he was having problems at school. I started taking him to school with Natasha and picking him up.

So, day-to-day life got started and we got into our routines. Natasha went out with her friends and I worked on my first book. For the most part, she tried to keep her friends and I apart. At this point, I didn't mind so much because I knew we had a cultural problem. Natasha and I were working on being married. It wasn't like she was some local prostitute.

Slowly, more people began to learn about us before we were married, which included the maid from Romania that she had known a long time. However, they were all in the trusted category along with her son,

Volodia, but we still needed to be careful. They were really great people, including her close friend that would visit us. I enjoyed them all.

I had a friend back in America going through an immigration situation with her son. She explained to me how it could go on for years and all the documentation of the relationship that they liked to see. So, even though I was not as much of a picture person as Natasha was, we began to take more photos with us together, as well as with family and friends. This was meant to show immigration that we had a real relationship when the time came.

So, we stayed focused when it came to getting married, went out when we could, and I worked. Natasha spent her time working out and being with friends. We still had some language barriers we were working through, though. Most of them were kind of funny, actually.

For example, she would come home and ask "Where did you go today?" This is not really a question to ask an American man. I knew that she was really just trying to ask me "What did I do today?" or "How did my day go?" However, asking an American man "Where did you go today?" kind of put me in the mode of "Fuck you! You can go somewhere, but I can't?!" We had many funny moments like this.

There was one other situation when she came home another day and basically asked the same question. I said, "Dead mouse." I needed a battery. Well, of course, I was talking about my computer mouse, but

Natasha really didn't get that and had a confused look on her face. When I reached into the plastic bag we were using for trash to show her the battery, she started yelling, "No! No! I don't want to see dead mouse!" Then I showed her the battery and said "Computer mouse." Our life was full of these moments and they were great.

In this respect, you have to embrace the chaos that is world travel and marrying into a different culture. Believe me, you can truly fall in love with it and there is an international language between a man and a woman. You don't have to know how to tell her you want her or how to figure out her body wants you. Mother Nature can do that just fine.

On the other hand, it was a peaceful time to come down from Saudi even though a war was brewing. People asked me many questions because I was the only one they knew from foreign war zones and, of course, nobody wanted to be involved in a war. It was odd being that guy they wanted to talk to about it. However, I didn't mind helping.

There were a few nights, though, I got to sleep by myself when my dreams got a bit out of control. Natasha would say that I spent the night screaming at people. One night, I even threw a punch into the headboard of the bed right above her head. After that, she opted for the couch on most of those occasions. Such is life, I guess.

We had to wait forever for the internet to get installed in our first apartment, so I would normally go across the street to the restaurant to work. Natasha always found it interesting that I could make friends anywhere in any language. Russians are not prone to talking to strangers. However, fuck it, I'm Irish.

That said, when we finally got internet installed, it was butt-ass cheap! By the time I was leaving Ukraine, they were doing fiber optic gigabit internet for $15 a month. What the hell was wrong with America? They overpriced everything.

There were many things to like about Ukraine, even though war was breaking out and many wanted to kill me. It was a beautiful place, in many respects, and most of the people were very nice. In that way, many countries I have been to are the same. There is good and bad everywhere.

People often asked me about how safe it was where I was living at any given time. They felt that their home was always a safe place and never saw the dangers right in front of them. I once read a 26-city list of the most dangerous places in the world and while there wasn't a single city from Syria on the list, there were three from America.

Even Natasha would ask how I knew someone, but I would sometimes get tired of sitting inside even when we got internet, so I would go and sit in a restaurant or bar somewhere and work. Of course, I would eventually take a break and talk to people. Not all were nice. Many tried to use you and con you.

Some would ask Natasha why she was protecting the American and she would explain that we were together. However, that was the life in Ukraine. I didn't blame them really. They were just a bunch of people trying to survive. From theft to piracy to sex shows and prostitution, they just wanted to make it.

Well, things were literally getting hotter by the day. I mean, the fucking place was on fire and they wanted the president's ass. They were going to do everything they could to get him, too. That was not a guy I wanted to be at that moment in time. However, so far, at that time, it was just Kiev.

It is amazing how something right in your backyard can seem so far away, though. Kiev was about 9 hours away from us by train. However, even though we were aware of it at the time, we were still pretty much ignoring it. The protests in the streets of Odesa right outside our window were, of course, a little different though.

Social media was beginning to blow up, of course. They were using any source they could and who could blame them? Rumors began to swirl that some of the shots taken were by American snipers. This was always possible. It's not like America hasn't done what they can to tear down, rebuild, or control the destruction of a government before. However, all of that said, America basically heard nothing at this point.

There finally came a day when it was time for President Yanukovych to run. And run he did. He got out and fled to Moscow. He and Putin had been buddied up for a long time. This ultimately put a dent in Putin's plans because he was really hoping to control Ukraine from the inside. That just wasn't going to work out.

Once Yanukovych left, it was much like Africa. It was a country without a government. Everyone began wondering what should they do now? They had to govern the place. It wasn't much different than the dog that catches the car and doesn't know what to do next.

Of course, over the next few months, in Ukraine and Russia, the usual activities began taking place. By this, I mean that people in the former president's family and on his team began to forget how to fly or drive. Many car accidents occurred that killed a number of people. Even the president's son ended up in a lake in Russia. Others thought they could fly from buildings, but failed to do so. They, of course, died also.

President Yanukovych himself seemed to avoid harm, but made INTERPOL's naughty list and would most likely never get out of Moscow. Maybe he and Snowden would become really good friends. Putin had made enough money to keep Yanukovych around, I guess, or wanted to have a bargaining chip, just in case.

Of course, I was not sure how great life was for Yanukovych those days when everyone around him was dying. It had to be a constant reminder that while he got rich ripping off Ukraine, his life was not worth much anymore. The Russians didn't care about how much money he had stolen. They cared about the fact that he was a fucking liability. He had to wonder if the kill orders came from Putin himself since he was former KGB.

So, the country began looking for a new president and planning a special election. Ultimately, at the top of this list, there was only one man. This was Petro Poroshenko. He was a rich businessman that had made his money in the chocolate industry. Unlike Donald Trump though, he had worked in government before.

Everybody basically felt that he was going to be the man and they started preparing for the special elections. Unlike America though, many in former USSR countries just let politics play out and didn't get involved. I know what you are thinking. I just told you how they ran a president out of the country, but really, most didn't get involved.

When it came to men like Poroshenko and Trump, I generally felt like they truly wanted to help their countries. I am not saying they were perfect, but anybody who loses money by walking away from a successful business to be a politician had to care more than the average person. Poroshenko was one of these men. He lost money walking into politics.

Not only did he lose money, but with the last president being run out of town, he was also going to be under the microscope. Not a place most people want to be. I mean, after all, it was not like all things were going to go perfectly, but he would be the one to blame when they didn't. It wasn't much different than some project management gigs I had taken.

So, Poroshenko began his campaign, if that is even what we should call it. I don't even remember anybody else or anybody talking of anyone else. There was another guy, but he was so far behind he could have been Al Gore. There was no chance of him taking the presidency from Poroshenko. All along, things continued to heat up because some still wanted to be Russian again.

Basically, the country decided that special elections would be held in May of that year, which was 2014. As it would turn out, May was going to be a busy month. You never know how a country will fall apart or a glass will break. We will talk more about that month later.

It was actually about this time that it became obvious that other countries were moving into the area to influence as they could. There was one thing about contractors and spooks. We didn't care what country you were from. We still tried to extend professional courtesy.

One day I was in my favorite microbrew place with Natasha when I received a call from a friend in the states. I knew he had cancer and had been fighting for a long time. He called to tell me he was going to die. He literally called from his death bed.

I had noticed another operator alone at a table over my shoulder in the mirror behind the bar and was keeping an eye on him. I took the time to say goodbye to my friend. After I hung up, Natasha and I had a shot of Irish whiskey in his honor. I knew the man behind me had been listening to my conversation, but it was not like it was confidential. When he got up to leave, he tapped me on the shoulder and said, “I am sorry for your loss.” That freaked out Natasha, but she was starting to learn the game was on.

Chapter 6: Hot Gets Hotter

Well, all eyes had been on Kiev, but focus suddenly changed. Russia had a committed logistical situation in the area of Crimea and it started to draw some real attention. This was a military consideration for Russia and they were not going to lose their foothold in the country. Which means, they took it. Putin basically said, "It used to be Russia and we just took it back. We didn't invade Ukraine."

Russia had always been strong when the USA president showed weakness, and Obama was a pussy. He started by trying to tell Putin to get out of Crimea and Putin basically said, "Fuck you." This went on for a while until Obama said "Ok, but no more than Crimea," just like a pussy will do. Obama did nothing, as usual.

Wars today have now really become black wars, off book, run by contractors for many reasons. This one was no different. There was a good reason for that.

In 1991, when the USSR collapsed, Ukraine held many nukes belonging to Russia. Russia, of course, wanted them back. Ukraine agreed to give them back, but not before getting a written agreement. The agreement basically was that if Russia ever moved against Ukraine, then Europe and America would come to Ukraine's defense. Obama screwed the pooch there.

Europe and America responded in their typically weak fashion by placing sanctions on Russia. Their idea was to fight guns with dollars against a former KGB agent. They seemed to have also forgotten about the fall of the USSR in 1991 and their ability to recover. Russia continued to move by using contractors to invade Eastern Ukraine.

In the end, or at least when we left, Eastern Ukraine had become a stalemate zone. Putin once publicly announced that he couldn't take Ukraine when he wanted to because of the international criminals that were there. Typical propaganda; he had contractors, but the rest of us were international criminals.

He knew what he was doing, though. Had he taken it and made it Russia, like Crimea, he would have had to support it. However, as long as it was a problem area not unlike Eastern Moldova, then it kept Ukraine from joining NATO. This is because you cannot have a problem with your neighbor if you are going to join. Putin knows this well.

The Eastern Ukraine problem could go on indefinitely. Putin would periodically fortify the border and run drills there. Ukraine talked about a wall, but as of when we left, it had not happened. Ukraine did sufficiently lack in military training, however. At one point early in the war in the east, Ukraine lined up soldiers behind their forwardmost advancing soldiers to shoot any of their own men that tried to run. Oddly enough, this is a common practice in many countries and shows the gravity of the problem when it comes to lack of training.

Ukraine went back to ordering a conscription process when we were still there, so many were worried. While others were fleeing to Ukraine to avoid their countries, Ukrainians started looking at fleeing their country to avoid their government. Maybe we should just all periodically change countries every now and again, like musical chairs, and call it even.

We began to think of Volodia, Natasha's son, and what his future would be like. He was still many years away from 18, so hopefully, he would never have to deal with the situation. However, it was becoming obvious that his country, that I was currently a resident of, was definitely beginning to change. Not necessarily for the better.

Since there wasn't much we could really do at that time, we continued to try to deal with daily life and not give it too much thought. You see, Odesa borders the Black Sea and that troubled area in Moldova that I talked about, and Crimea was very close and now occupied territory. That put Odesa, the largest port on the Black Sea, in a perfect hammer and anvil situation from a military point of view.

Reporters were beginning to come into Ukraine from all areas and generally getting their asses in trouble. Many reporters like to push the limits to get a story. Often, they went to the Crimean border. Russia isn't too keen on shit like that and many were detained and almost shot. It was not a good approach to getting a story.

The Ukraine women took another approach that was kind of interesting. They were basically at the border in their hottest outfits chanting, "No pussy for Ivan." So, the Russian soldiers knew they weren't getting laid by the most beautiful girls in the world. It was an interesting tactic, but I guess you use whatever you have at hand, so to speak.

In the meantime, Natasha and I continued to finalize plans to get married. Ukraine was not a big place for church weddings and it really wouldn't have mattered if it was. Natasha and I had both been married before, and in Ukraine, you basically only got one marriage in a church. This was a place that was extremely anti-divorce. They make it hard and you only get one shot in the church.

However, it was our wedding, so it was important to us. We didn't have as many details to work out as it would have been to get married in the USA or somewhere else, but there were still things to do. I had already purchased some very nice high heel shoes for Natasha that she enjoyed wearing, but we were still short a few items and she had to figure out who was really going to know about our secret marriage. This would be the first time that either one of us had to pull off a secret marriage and we were still wondering about the ramifications. So, there was a lot of conversation between the two of us about it.

Our needs for the wedding were pretty basic. We needed to make sure we had Volodia that day, a dress for Natasha, and of course, flowers. She was also going to need someone to come over and get her all dolled up the day of the wedding, but she already had a line on this.

That said, for men, it is always a lot easier. Polish your boots and throw on a jacket, tie, and pants; done deal. We did look at places we would like to eat afterward. Natasha went through tons of flowers and decided what she wanted. Then we spent time in the mall shopping for a dress and trying on many different ones. As I said, she always needed to be the most beautiful thing on the planet. Once again, she succeeded.

The marriage office was by the opera house in Odesa near the port. It was a nice place for photos, so we had that in mind. Eventually, the day came, and Natasha started getting ready. I wouldn't need as much time, so I ran out for her flowers. Of course, I also got a beer because sitting around watching women pretty themselves up was not really a spectator sport for me. I also picked up Volodia.

Picking up Volodia at this time was always a bit tricky. After all, we were getting married in secret and Volodia's father, Sergey, was one of the reasons why. He was a diehard communist and Putin fan, so letting Volodia run off with an American whom his ex-wife was marrying was not high on his list. Lucky for us, Sergey basically only left his apartment for 2 hours a month on one day.

Natasha finally got finished up with her hair and nails and I got cleaned up and dressed. Her mother showed up and her friend was meeting us at the marriage office. We went to the opera house to start our day of photos. There was everything from a dove release to photos of Natasha by the trees. Then we went to the marriage office.

It was an interesting ceremony, but not much different than in other countries. There were more photos before, during, and after the wedding. The ceremony consisted of vows standing on cloth runners that we would keep so that we would always know where we were standing, I guess. Additionally, there was a candle ceremony, a cake, mom gave away Natasha, and then the signing of documents. Had this happened a few years earlier, I would have been getting married in front of a photo of Stalin, which really would have helped me in America. Not!

We went out to eat and have drinks and actually hit two places. When people get married in Ukraine, to some extent, the whole city is involved. First, we had pizza and then went to our favorite microbrew place. It was right next to a very nice park and it was a great day. All in all, we got back to the apartment pretty tired, but, of course, happy.

We got married on March 15, 2014, so there was still half of the month left, and Volodia had spring break coming up. With Natasha, traveling was always a challenge because a Ukrainian passport was not one of the most favored items in the world. We wanted to go somewhere for our honeymoon and wanted to take Volodia with us since he would be on his break. I left the planning of this up to Natasha.

Volodia had other problems when it came to traveling. While Americans can get passports at a very young age, Volodia couldn't get one until he was 16. So, he was in Natasha's passport. However, in order to leave the country, we needed signed documents from his father. This was always a pain in the ass because Natasha didn't like dealing with him and was still trying to keep our new marriage quiet.

So, I went back to work on my book and Natasha was off on her mission to find us a place to go and get Volodia out of the country for a week. She was a master at finding great deals on all-inclusive trips. Most people would search them out on their own, but she actually had success with travel agents. This was normally how she traveled, which is why she always ended up in Russian-friendly places. I traveled for business, mostly, and went where the work was, regardless of the challenges.

Well, in this case, once again, she was successful. She found an all-inclusive resort in Egypt by the Red Sea. She was always thrifty with our money, but I was deeply impressed by what she had found this time. She managed to get a week for all three of us for $1750 or something, and that included airfare. The resort even had a water park with slides, which was something Volodia really liked.

Now, Natasha just had to fight with the ex to try to get documents for her son to leave the country. I mean, this guy was more than useless. Their family owned rental properties in Odesa and that was how they made their money. He only had an eighth-grade education and spent all his time playing video games. He would ask Volodia, "Why do you need school? When I need a shirt, they give me one. When I need food, they give it to me." Purely communist thinking.

How Natasha ever married him or why, I have no idea. That was not her thinking. She went to law school and entered the Ministry of Interior of Ukraine. Hell, when Volodia was born, his father didn't even want him to know that he was his father. Volodia's grandmother was the one that told Sergey to get a divorce from Natasha when she got pregnant, so she couldn't get anything from the family.

It was an odd situation I never truly understood. However, it was clear that Volodia's grandmother was definitely the head of the family. She called all the shots. Maybe she had paid for the buildings that were creating their income. I just didn't know.

Egypt was a visa country for all of us, but like Turkey, they make it really easy. You just hop in the visa line when you get there, give them the $10 per person or whatever the hell it was, and they put the visa in your passport. After that, you went to customs just like you knew what you were doing all along.

We got through customs, found all our shit, and headed for a taxi. Egypt was going through an interesting time in March of 2014, but then I thought Egypt was always going through interesting times. They were basically on track for a change of leadership also and their election was tracking pretty close to Ukraine's election dates.

Cairo was getting nastier and the military was being attacked. Hamas was declared a terrorist organization, and 528 people were sentenced to death. It was the largest mass death sentence handed down anywhere in the world. However, we were on our honeymoon, so, fuck it. We got a taxi and went to the resort.

The resort was at the southern tip of the Sinai Peninsula in the city known as Sharm el-Sheikh. It was a really nice resort and Natasha had done a good job. Of course, when we checked in, they did the typical "Can I see your passport?" routine.

Hotels often wanted to hold your passport. I didn't know why people agreed to that. I was always willing to show my passport and give them a copy, but I was not turning loose of my passport. Well, this time was no different. After Natasha had produced hers, I showed the guy behind the counter mine. He quickly said "American? Americans don't come here." I told him one was here now and I would be hanging onto my passport.

We went to the room, got unpacked, and Volodia settled in with the internet. Natasha and I had a resort orientation to go to. We were still kind of early for it, so we left Volodia unattended and went to the bar. It was a place where we really didn't have to worry about him because there were a lot of children and decent security. That was a nice perk.

Eventually, we got to the orientation and they told the small group there about the resort and all the additional cool shit you could do. Natasha got a pamphlet to look over. It took her a bit longer because it was in English. We went to the bar so she could look at it and figure out what she wanted to do over the week. I was usually pretty flexible. With all the traveling I had done, not moving was often what I liked to do.

I really wanted to see the pyramids, but that was in Cairo, so there would be no taking Natasha and Volodia, given the security issues there. I didn't mind risking my own ass and never had, but I wasn't going to risk theirs. So, the next day, I pulled a guy aside and asked him about going. He said nobody was going in, especially Americans. I asked about getting a private plane and flying in myself. He said I could, but the pilot would probably leave me behind to get out on my own. That excursion was off the table of options.

For the most part, we stayed at the resort, but there were a few things we did. There were kids' activities, so Natasha and I were able to go places alone with Volodia supervised. It was really a great resort.

I noticed a Hard Rock Café just down the street. Natasha had never been to one, so I took her down there. We got something to eat and some drinks and, of course, looked at all the pieces on the walls. I did have to spend some time with Natasha explaining the concept of the Hard Rock Café, but she did seem to enjoy it.

On most days, Natasha got some sun while Volodia was in the pool on the slides, and I got a little work done. I never completely took time off. As I may have mentioned, Natasha was not the most coordinated person in the world, so she couldn't ride a bike or swim. However, the pool was shallow enough she could get on the slides with Volodia and not risk drowning.

In the evening, we would spend time on the outdoor deck or in the bar meeting other people. While there were wars breaking out all over the place, it was not obvious here. Russians, Ukrainians, Muslims, and their children were all able to get along just fine, and most did not agree with what was happening in the world. It was obvious it took governments to really fuck things up.

While at the resort, I introduced Natasha to the concept of massage. Specifically, Thai massage. She had never really had massages before and I was feeling a little beat. Since there was a spa, I decided we should take advantage of it.

We signed up for massages and I got a small, little Thai girl while Natasha got an Arabic girl. It would take several months before Natasha finally got the Thai concept, but this began her learning process. She was able to get rubbed down and hop in the hot water while waiting for me to finish and was pretty happy with her spa experience for the first time.

We had two massages during that week. By the end of the second one, Natasha was saying that she should have had the Thai girl because her "monkey" was useless. Her broken English always made me smile and fall in love with her even more. For that reason, I guess, I found it hard to correct her many times. At least she was willing to try new experiences. In fact, after that week, she was on a mission to find us good massages in Ukraine.

Natasha still had a couple of big events planned for the week though. The first one on the list was to jump on a couple of quads and ride out through the desert. This would be a day-long trip and was really nice.

Initially, we rode a pretty good distance, stopping occasionally for photos or just to take a break. Natasha did pretty good and didn't manage to throw Volodia off, so that was a good thing. Of course, a desert is basically a desert. This meant it was full of sand and rock.

Eventually, we got to the camp that we would be eating at that night. At that point, we traded the quads for camels and went a bit further into the mountains. We got back about the time dinner was getting ready. It was a pretty nice camp. We had tents to climb into with tables for our food and lots of cushions. There was also a little stage that had been set up with a Russian belly dancer.

I can't remember exactly what we ate, but it was traditional, so I am sure it probably consisted of lamb and rice. We continued to watch the show and finished dinner. Later, they had a lesson in astrology. We were, of course, a long way from city lights and it was a clear night, so the stars were great.

A telescope had been set up for stargazing, which was a new experience for Natasha and Volodia. I always liked them having new experiences. Life in Ukraine, with the fall of the USSR and now the war, was not always the best, so these moments were nice. The final thing to do that day was wander our asses all the way back through the desert, only this time at night. We did, and all was well as we returned to the resort.

It was still early when we got back. The kids were working on a show to perform at the end of the week, so Volodia ran off with them. Natasha and I just relaxed and took in the night sky. We also continued conversations with our new friends from all over the world while the kids got ready for their performance. Since Volodia was actively engaged in dance in Ukraine, these shows were something that he really enjoyed.

The week was coming to an end, but we had just a few more things on Natasha's list. There was a beach nearby, so we would spend time there when Natasha got tired of the pool. It was nice and clean with benches to use and a bar with internet. Volodia even liked the change of scenery.

The last day there, Natasha had us booked on a boat tour of the nearby Red Sea. We went by a dive shop before the boat, so we could get Natasha and Volodia outfitted to paddle around without sinking. I would have liked to have gotten some gear and gone diving, but we were flying the next day, so I couldn't. It was still a nice ride though.

The Red Sea was very clear and certainly beautiful. I was glad to have gotten to see it. There was a photographer on board, so Natasha and I got some more photos. It was a very nice time and we didn't really have to think of Ukraine at all, except for the Ukrainian woman that Natasha met.

Among other things, I was a deep wreck and technical scuba instructor, so wrecks were something I fancied. There were actually a few on this tour that weren't completely underwater, so it was kind of a neat thing to see. Near one of them, they anchored up so that people could get out and swim. I got Natasha and Volodia all geared up so they could get in the water safely and watched them from the boat to make sure they didn't get in trouble while they paddled around. Once again, they had fun.

Natasha and Volodia finished up with their swimming adventure and we got them back on the boat. Once their gear was stripped off, we settled in for a nice sunset cruise back to the dock. Since we were leaving the next day and Volodia had his show to put on with the other kids, we got back to the resort and packed up for the next day.

Afterward, we went down by the pool where the resort had set up a small stage for the kids.

The kids put on a nice performance and, once again, it was an example of how the world could get along with each other if just left alone. A Russian named Alex came over to our table with his wife and some homemade Vodka. They were quick to tell Natasha and I that they didn't believe in or agree with the war. They even went so far as to write a note about it and sign it so that we could take photos of it and asked us to post it on Facebook.

It is moments like this that truly teach people about the world. I think more people should experience them because their views would certainly change. Don't get me wrong, I know there are people in the world that want war and are truly evil and I am willing to fight them. However, from my personal experience, the common person in just about any country just wants to live life.

Well, we got up the next morning ready to go home, even though we knew it wasn't one of the best places in the world and was getting worse. We said goodbye to everyone, went to the airport, and boarded the plane. My mind always tends to shift back into tactical mode when flying into areas like Ukraine and this time was no different. It was time for a beer.

Chapter 7: Putin Stare Down

As more than expected, things continued to heat up all over Ukraine, but eyes were really on the east. There were, of course, rumors of Russia being behind it, and ultimately, Russian soldiers were captured in time. However, the country was becoming divided by those that wanted to be Russian again, like Crimea, and those that wanted to remain Ukrainian.

Protests were happening all over Ukraine, including Odesa, which was right outside our window. Some were peaceful and some were not. Armed combat began to take place in the east where government buildings and arms were seized by rebels. However, it appeared that even the rebels were not unified about who was in control. There were multiple separatist groups claiming they were the rightful new government of their regions.

The government continued to fall apart with members of the police choosing sides and leaving their posts and duties behind. Even judges had resigned due to the corruption that they were seeing and took up arms. Whether these people knew they were being manipulated or truly wanted a civil war was hard to tell.

Many of the rebels were experienced veterans of the former USSR and Ukrainian wars, but there was still a rather large group of civilians.

As usually happens, there were many un-organized skirmishes. After each one, many of the civilians ran off realizing they didn't have the stomach for a real fight. The rebels were able to seize military vehicles and arms from Ukrainian troops, so they weren't a bunch of farmers with pitchforks.

Activity generally revolved around Donetsk and Luhansk, but, as it was a port city, Mariupol also experienced a lot of tension. These areas and the surrounding small cities were all home to the people there, whether pro-Russia or pro-Ukraine. For this reason, most did not leave and united within their respective groups as best as they could. Of course, bodies began popping up, prisoners were being taken and released, and many were being beaten or killed for speaking anything other than Russian.

Additionally, media outlets all over the country were being taken so that rebels could spread their propaganda. Many hoped that Russia would come in and annex them as they did Crimea, but Crimea served a different purpose. Both sides stupidly hoped for help from the UN, but Russia had veto authority in the UN for any kind of decision like that. Some started to cross the border into Russia looking for a new home and ending up in shabby camps quickly erected.

All in all, the interim Ukrainian president was having a hell of a time. He tried to put together a few different forces to kill the rebellion and re-take the east, but with very little success. Ukrainians seriously lacked in the area of military training and then there was the issue of mixed emotions that typically comes with a civil war.

Basically, for the most part, the two sides just took turns killing each other and trading land. During these days, the lines moved often, eventually stabilizing. Most of the targets being hit were government buildings that they wanted to re-take or armed roadblocks to restrict movement. Of course, anything transporting troops caught bullets whether it was air or ground transport.

Monitors from all over the world, especially Europe, began making trips into the area and some were captured. Many of the military units were not very big. Some of the largest were only a couple hundred people, but they were able to get results in controlling areas and seizing weapons and ammo. Molotov cocktails were used, but they had always been an effective improvised weapon.

The monitors brought with them the attention of the world, of course. I never really understood the use of these monitors. I always saw them as more like reporters who were counting on some kind of diplomatic status or something. These groups would be the perfect place to insert militants, so, of course, they were going to be looked at closely.

Mostly, the civilians, regardless of what side they were on, managed to stay out of the way even though their homes and cities were being destroyed. There was the occasional death, but it was to be expected. At least civilians were not being deliberately targeted. They weren't happy by any means, and supplies became an issue. It was not like they were sitting around drinking champagne and cognac watching the war.

With all the back and forth of trading land, it became obvious that no side was getting anywhere quick. The rebels did resent the damage to their area even though they had fucking started the situation. I guess they didn't ever expect Ukraine to fight back or were just plain stupid. Odesa was considered one of the most beautiful places in Ukraine, so the rebels would say that they were not stopping until they destroyed Odesa as payback.

Natasha maintained that Putin did not want Ukraine, but I was not as convinced. However, Natasha, like many, had mixed emotions. She had family in Crimea and in Russia, but Odesa had been home for four generations. It was a confusing situation for many. Worried, she tried more and more to make me look Russian. I continued to insist that if I died it would be in my cowboy boots and hat.

Most of the Ukrainians I knew were very nice to me and accepting, much like the people we had met in Egypt. That said, our marriage was still a secret, and she kept me away from most people that she knew. Her family, other than those at the wedding, still knew nothing. I did begin to see instances where people with cameras began getting the shit beat out of them for taking the wrong photos. This has always been a sign of changing times.

To a certain extent, as an American in situations like this, you begin to feel like the guy tied to a tree butt-ass naked with a pork chop around your neck in a pack of hungry wolves. You know that it could get bad anytime, but there is not much you can do about it. I did maintain a relationship with our US immigration lawyer, but that didn't seem like it was going to move faster in the near future.

As it turned out, immigration was about to become a problem for both of us, which would cause some meetings that Natasha had been putting off. You see, as an American at that time, you could be in Ukraine for up to 6 months a year, but only for 90 days at a time. I had been calculating my in-and-out time since 2013 when I first arrived there and realized I was about to have a problem. Natasha couldn't go to America, so I would have to figure out how to stay in Ukraine. At least we were married, so we hoped that would make things a little easier.

Once again, we had to tap Natasha's former lawyer skills from the Ministry of Interior of Ukraine. She had gone to elementary school with some people who now ran a notary office. This was our first stop. In typical Ukraine fashion, their names were Nicoli, his wife was named Natasha, and his sister was named Natasha. Can you figure out what the popular name in Ukraine was?

Now, you must understand that notaries in Ukraine were nothing like America. In America, these people just stamped docs and said he was the guy that signed it. In Ukraine, they were more like paralegals where it was their ass on the line if they screwed something up. This made them a very good source of information.

So, we drank some cognac with fruit and chocolate, talked about life, and explained how my Natasha had married an American and what our problems were. Once again, there was a twist. I had to leave the country and come back in, which would have been a really great thing to know before we left for Egypt. Welcome to world travel and shit. The upside was I didn't have to go all the way back to America and Ukraine was smaller than Texas.

We determined, after some more phone calls, that I could go to the embassy in Moldova for a different kind of visa to re-enter Ukraine. Apparently, I couldn't obtain residency on a tourist visa. I had to leave and get a different visa to return on for residency. I guess every country has a certain amount of bullshit. It was still easier than America.

This was a former USSR country that still had USSR ways. Much like the USA uses your driver's license and address to track you down when they want to bust you for something, the USSR countries had you "register" at a residence. Unlike the USA, however, you had to actually own the place where you were registered. This meant that a rental apartment was not going to work. Of course, I was not in a hurry to buy anything in Ukraine with war breaking out.

This was another time where we were going to have to learn the law and then what really happens. We found out that many people would just buy a cubic meter brick structure outside of the city that had an address and register there and then they would rent inside the city. In fact, the cop living right next door to us, at the time, had done just that. On top of that, I found out that Natasha's mother was renting out the place that she owned and was registered at, but was living with Lena, Natasha's sister.

This seemed to be a rather typical arrangement, especially for the older parents. Ukraine was much more family-oriented than in America. They took care of each other. In fact, Natasha's mother and father had gotten divorced years earlier, but they both lived with Lena, her husband, and daughter. I couldn't help but think that America could learn a great deal from Ukrainian people.

So, this left us with only two choices that would cause the afore-mentioned introductions that Natasha was trying to avoid as long as possible. I either had to register at the house my mother-in-law was renting out or at Lena's, who didn't even know we were married. Either way, one of them would have to go to the registration office and sign for me. To make matters worse, Lena didn't actually own the house because it was in her husband's name.

It even turned out that getting "registered" was nothing like using a friend's address in America for a driver's license. Once you were registered at a property, if the people that owned it decided to sell, they needed your signature. So, you were definitely becoming part of the family. This required a few drinks.

As it turned out, Lena's husband helped out at a disco and they called Natasha to see if she wanted to go because we could get in for free. That gave us the opportunity to break the news that Natasha and I had married. Neither her, her mom, nor Volodia had told anyone. In Natasha's typical graceful fashion, when we got there that night, she still didn't say shit. She just waited for Lena to look down and see that we were both wearing wedding rings.

At that point, Natasha had some explaining to do. Lena let out a squeal, but it was a happy one, which of course, is always good coming from a woman in any language. I let them talk while I got through my beer and a few more. Sergey, Lena's husband, was a nice guy, and while we didn't fully talk each other's language, we got along pretty good.

It was a nice night and we got over the hurdle and fear Natasha had of her sister finding out she had married an American living in Ukraine. We still hadn't addressed registration, but I knew that Natasha would handle it. She was very aware of our timetable, so I left it completely up to her.

Over the course of the next week or so, we spent time at Lena's house. I got to meet the rest of the family and the hairless cat mystery. Those things were just fucking weird, I must say. Oh well, wasn't my cat, so who gives a fuck?

Natasha's father was a highly decorated member of the former USSR Army during the Cold War, but for many years, he had been bed-ridden from a stroke and unable to speak. This was extremely unfortunate since he used to be a Russian/English translator. However, he was still pretty alert.

When Natasha and I went into his room to finally meet him, she tried to tell him that I was Russian. He barely ever rose up out of bed, but did on that occasion. I could see the old Soviet Union in his eyes and respected him for that. He was such a strong man even laying there trying to die. He simply looked me dead in the eyes and then looked at Natasha and said, "That's no Poruski." He figured out real quick Natasha had married an American. He was a good man.

Natasha saw a weak, frail, drunken man in her father. She saw disappointment and pain. She saw an old war and an old time and a bad life, but I saw something different. I saw a forgotten hero of a fallen country that had been left behind to die in the most miserable way possible for a guy that had seen the world and had been respected. I saw a brother.

As I was told the story, when he first had his stroke, which was about 7 years earlier at that time, they tried rehab to get him going again. He eventually fell out of bed trying to go to the shitter, when he could still do that kind of stuff and broke his hip or pelvis or something. Whatever it was he broke; he went down for the count. The poor guy could only lie in bed and watch TV after that. When the family tried to explain that they had no more money to send him to rehab since the country had fallen, he simply pointed at the only thing he had left, which was his TV, to suggest it be sold for the money. He pointed because he was unable to speak, but he was going to fight to the end, and I respected that.

As I have tried to detail here, many individuals think that people like Russians and Americans should be enemies, especially combatants, but it just isn't true. Yes, we were trained to hate each other, but from my experience, we just saw a brother from another mother. We were operators and knew the pain that the other side went through even though we were taught to envision the other side and kill them. Would we have? Unfortunately, yes. Should we have? That's a whole different question, especially when you married them.

At the deepest depths of life, soul searching, and, I guess, adventure, you find questions you never really knew you had. You have arrived at the crossroads of a situation you were never trained for when you thought you were trained for everything. You have to sort feelings from tactical responses all while being in a country where you were alone, not welcomed by everyone, and didn't speak the language fluently. Words can't describe the experience and what you feel even though I am writing this book to try to tell you about it.

In that respect, I may as well have been on the dark side of the moon alone. I remembered some of my deepest, darkest, coldest water dives when I was alone. I knew it was all me or I was dead. You knew not to fuck it up because there would be nobody that could save you. Ever. Yet, with all that in the back of my head, I knew I loved Natasha and Volodia. I just kept moving forward. I had no other excuse for my actions at the time than that. It was simple. I wouldn't let them die if I could do something about it.

There was always a part of me that wanted to go to places and do things that nobody else had ever done. My grandparents, who I have already said were basically my parents, knew this. They never tried once to hold me back even though I know it concerned them more than a time or two. Maybe it was a lucky thing for me that they were dead by the time this adventure came along or maybe I made it because they continued to watch over me.

No matter where I was in the world, I made sure I was there the moment both of them passed away. They both died with honor and basically both said the same thing: "What are you doing here?" "Don't you have something to do?" "Weren't you in some other part of the world?" "Get back to your life." When granddad died his last words, when he looked at everyone, was that I would be the best of all the family. Some called that my curse.

I'm the last of the family name. I have no children of my own nor will I ever. It was something I gave up for the life I led. I always knew if I was going to be "that guy" that it would come at a cost. I can't work in an office or live in the house with the white picket fence and barefoot and pregnant wife that bitches at me all day. I marry countesses like Natasha.

I know what I have paid for my lifestyle. I drink like James Bond and have all the same quirks of trying to overcome the last gig while getting ready for the next one. I guess Natasha saw her father in me because it always seems that girls fall in love with their dad. Unfortunately, they usually hate their dad also.

Well, the day came when Natasha and I would need to make a border run. This would seem easy enough because Moldova was only 30km from Odesa, but there was a problem. There's always a problem during a war. We had to cross a Russian held conflict zone right between Ukraine and Moldova. They called it Transnistria. For lack of a Russian keyboard, I can´t spell it right here. As the name may sound, it was a transit zone or a place you had to cross.

The bitch was, as I said, it was Russian-occupied territory. This meant that an American married to a Russian born Ukrainian in the middle of a Ukrainian war and Russian occupation may have drawn attention, to say the least. The guards fell on me like a high school boy falls on wet panties on prom night. I was staring at a very nasty prison again and a lot of guns.

See, when you have traveled to all these countries, you learn a bit of many languages, but I couldn't read Russian for shit. Of course, Natasha didn't think of that or the fact that I didn't know the laws as we pulled into a conflict zone. Well, the sign on the door basically said to not go to the shitter when the train was stopped. On top of that, I came out of the bathroom like I owned the place, asking Natasha, "What did you do with the fucking champagne?" I was greeted nicely by AK-47's. This is where a man with my background evaluates a situation quickly.

Natasha immediately jumped up and basically said, "What the fuck?!" They asked why I was in the bathroom, and she turned lawyer on them. She informed them that I didn't read Russian and asked them why it was even open. She backed them off and I got 3 automatic weapons off my head, but they seriously went through our paperwork.

I knew I had two days left, but this first fucker came over and said I had blown out of my visa, which was bullshit. Natasha knew the situation, too, and said, basically, that was not the case. Another person finally came over with a computer, which now meant that my passport was going to be tracked, said I was good to go, but that they would be waiting for me to come back. This was always something that you want to hear in an illegally held, unrecognized combat zone. It was time for more booze.

After we pulled out of border control for that area, I began to realize we were on the slowest fucking train known to man. It was just wooden seats, but it took the same amount of time to get to Chisinau, Moldova as it took to go from Odesa to Kiev. This basically meant about a 9-hour trip.

I hadn't planned for that. We had left in a bit of a hurry, so we didn't have a place to stay when we got there. I was not in a hurry to tell Natasha that. We finally arrived around midnight and walked out of the train station. I looked around and wondered what the new hell would be. I didn't have to wonder for long.

Natasha quickly asked, "Do we have a hotel?" I said, "No." Then she asked, "Do we have a plan?" I said, "No." Then she was like, "You brought us to a foreign country in the middle of the night where neither of us knows the language without a plan?!" I said, "Yes, I did. Trust me." Then I said, "See, there's a bar. Let's get a beer."

Of course, in the process of walking a block or so to the bar, I got my ass chewed for being in the middle of Moldova with no fucking plan; but, this was usual activity for me. We went in and I got a beer for me and a glass of wine for Natasha to shut her cakehole. In a little Russian and a little English, I talked to the bartender and got us a hotel. He walked me outside and around the corner and pointed right at it and said, "There you go." I went back in and told Natasha we had a place and, of course, she asked, "How the hell did you do that?"

Well, we went to the embassy the next day and found out that there were "special" days we could apply for a visa. Our special day was a holiday, so they would only be open for a half day. Basically, we had to fuck off 2 days. I shot rubber bands at the ceiling, banged Natasha, worked and drank, just waiting and figuring out how getting back in the country was going to go. Fuck it. There was nothing I could do at the time. I had another.

As it turned out, when we finally got to our "special" day, it was probably a good thing it was a holiday. All they wanted was to get everyone processed and go get drunk. We did have some more hoops to jump through, like getting medical insurance, which I think cost me $15, fuck Obamacare, and we sat outside with a 12-pack waiting for them to call us in for our completed paperwork. We talked to people, of course. Some were Americans straight out of Crimea just trying to survive like the rest of us.

In the end, it went pretty smoothly for an unplanned border crossing through Russian-held territory. However, we got what we needed. We then grabbed another night of sleep and jumped the train again. At least this one had a bar. When there is a bar, you can drink on the train, but if there is no bar, you can't. This was something Natasha failed to mention before I had AK-47's pointed at my head asking where the champagne was the last time. No, there was no bar on the first train. All in all, we got home safely and prepared for the next adventure.

Photo Section

Natasha in our First Ukrainian Apartment

Natasha and Volodia on our Wedding Day

Natasha on our Wedding Day

Natasha and Her Mother Tatiana with Volodia and I in the Background in the Mirror

Natasha and I

The New Family

Me in our Favorite Microbrewery

Signing our Lives Away

Wedding - Russians Wear their Rings on the Right Hand

Wedding Ceremony

Natasha and I with Volodia and Natasha's Mom and Her Friend Katja at the Wedding

Natasha Out for Dinner

Typical Odesa, Ukraine Club

Odesa, Ukraine Opera House

Snow could really stack up in Odesa. This car was only parked about 3 hours.

World's Greatest Muscles and Our Favorite Microbrewery

Natasha and her Cousin

Natasha at the Opera House

Statues by the Port

Port Area with Natasha

The Odesa Port

Our Hotel Before Our First Apartment

Odesa Architecture

Unisex Bathroom Sign in Odesa

Natasha Modeling my Hat

The Family

Volodia Not So Shy

Natasha Being Cute and Funny

Natasha Just Being Beautiful

Want Vodka?

Going to Kiev

Going to Kiev

More Liqour Packages

Ancient Artifacts Near Odesa Port

Natasha on Train to Kiev

Kiev

Natasha at Titanic Exhibit

Things Heat Up In 2013

Me with Former Russian
Commando on Train to Kiev

Kiev

Unlike America, You Can Jump Out of Windows in Ukraine

Kiev

US Embassy in Kiev

Kiev

Euromaidan Preparing for 2013 Kiev Protests

Kiev

Kiev

Natasha at the Microbrew After Our Wedding

Natasha and I on the Red Sea for Our Honeymoon

Natasha and I on the Red Sea for Our Honeymoon

Natasha and I on the Red Sea

Natasha and I on the Red Sea

Natasha on the Red Sea Although She Really Couldn't Swim

Me on the Red Sea

Map on the Beach in Egypt

Hazy Egyptian Day

The Mountains in Egypt

Hotel Bar in Egypt

Hard Rock in Egypt

Camel Parking in Egypt

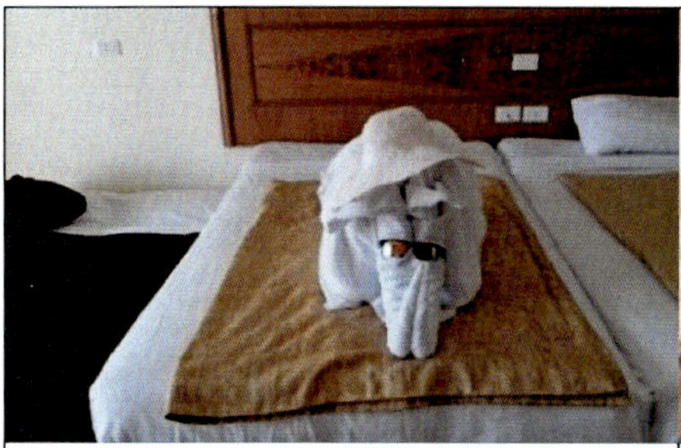
Egyptian Pranksters

Natasha Ready for Egyptian Desert

Natasha on 4-Wheeler with Volodia

Desert Caravan in Egypt

Egyptian Desert

Egyptian that Liked my Hat

Natasha on a Camel in Egypt

Switching from 4 wheelers to Camels for the Next Leg in Egypt

Egyptian Sunset

Egyptian Sunset

Photo Fun in Egypt

Photo Fun in Egypt

Photo Fun in Egypt

Making Camp for the Evening in Egypt

Dinner Entertainment in Egypt

Natasha Wanting to Pilot the Red Sea

Red Sea

Russian Note to Natasha and I

Natasha and I at Egyptian Bar

Red Sea

Natasha and I on the Red Sea

Red Sea

Saying Goodbye in Egypt

Saying Goodbye in Egypt

Drawing of Me by Venezuelan

Lena (Natasha's Sister) and I at the Black Sea in Odesa

Natasha and Her Sister Lena

Tatiana and Volodia

Natasha and Her Sister Lena

Natasha and Her Sister Lena

Natasha in Ecuador

Natasha, Me, and Mom

Chișinău

Me in Chișinău

Chișinău

Natasha in Chișinău

Natasha in Chișinău

Our Microbrew Place in Odesa

Katja and Natasha at the Disco

Crossing Transnistria Moldova

Chișinău

My Bird Friend in Odesa

Odesa Massacre

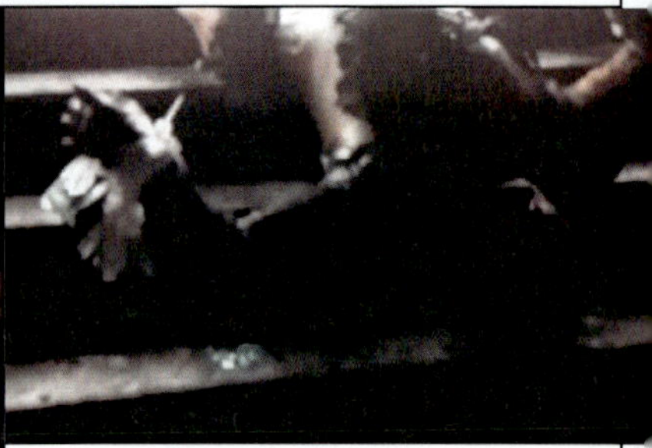
Odesa Massacre

Odesa Massacre

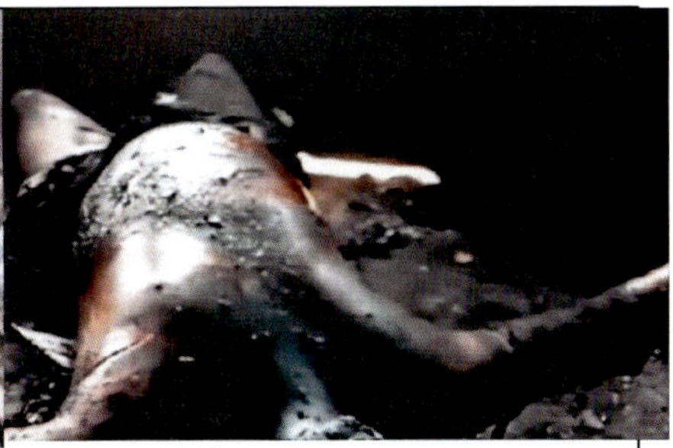

Odesa Massacre

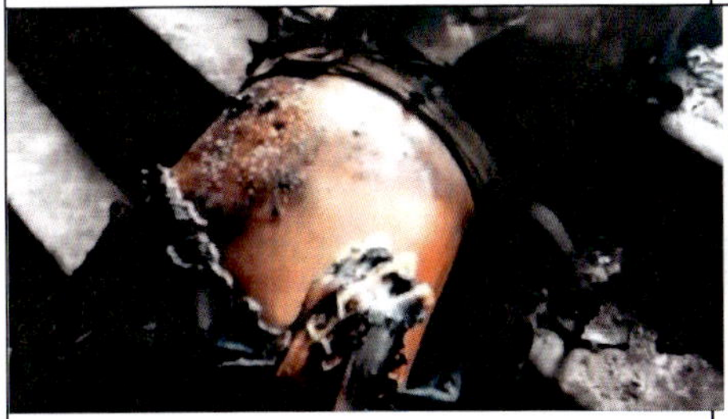

Odesa Massacre

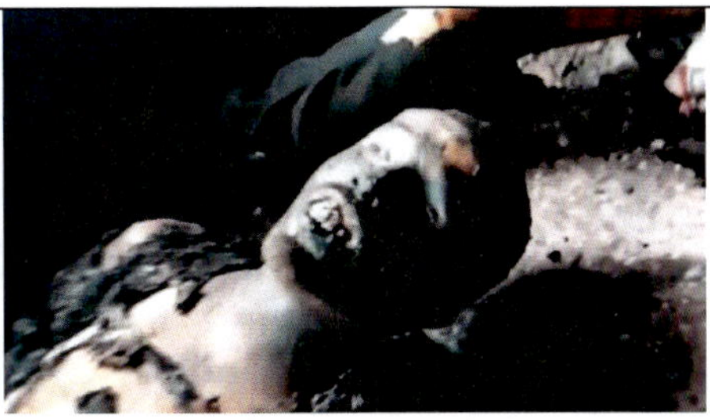

Odesa Massacre

Odesa Massacre

Odesa Massacre

Odesa Massacre

Odesa Massacre

Odesa Massacre

Odesa Massacre

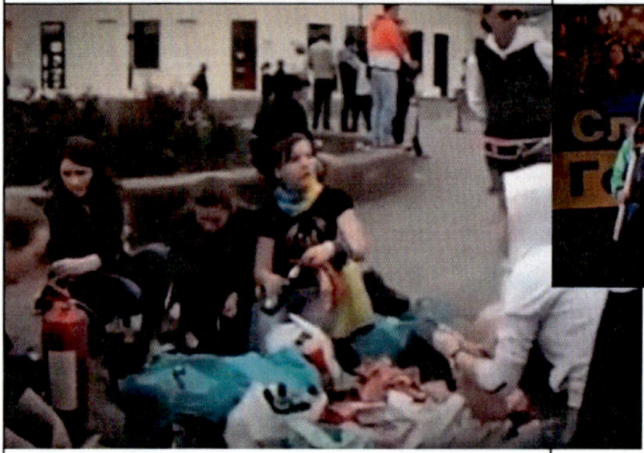

Odesa Massacre – Making Molotov Cocktails Where I Shopped That Morning

Odesa Massacre

Volodia and His Cousin Tatiana at the Black Sea Beach in Odesa

Me on the Black Sea Beach in Odesa

Natasha in Ecuador Shortly After We Arrived

Natasha in Ecuador Shortly After We Arrived

Natasha in Ecuador

Natasha in Ecuador

Natasha in Ecuador

Natasha in Ecuador

Natasha in Ecuador

Natasha in Ecuador

Natasha in Ecuador

Natasha in Ecuador

Natasha in Ecuador

Natasha in Ecuador

Natasha in Ecuador

Natasha in Ecuador

Chapter 8: Shit Gets Real

Generally speaking, we knew May was going to be a bit of a fuck-up. We could feel the storm coming. It's funny how you feel these things. Maybe it is in our natural protective makeup. All creatures have instincts, and that includes humans. We are nothing more than animals.

We didn't have to wait long. As we walked around the city in the first couple of days of that month, we could see things were being staged for an "event" of some kind. I didn't think Bon Jovi was coming to town. There was wood stacked up everywhere. Someone was preparing to keep the place "warm."

Both Natasha and her mother had birthdays in May. In fact, they were only two days apart, which was kind of weird. Her mother was May 2nd and Natasha was May 4th. On the 2nd, I woke up early, as usual.

We thought we lived in a pretty safe area because we were in a great part of the city and right next to the police station. It is amazing what you really don't know. A feeling of safety is complete bullshit. It really is.

That day, I started by going to the store while Natasha quietly slept. Since it was her mother's birthday, I wanted to start the day off the best way I knew how. I went to buy kielbasa and champagne just because

Natasha and I liked it. She was happy that morning and it was nice. It's funny how quickly things can change.

Her mother's birthday probably saved our lives. We left to go spend time with the family for Mom's birthday at Lena's and Sergey's home. The houses there were built like motherfucking compounds. It wasn't long before Sergey came to me with his phone and showed me the city on fire. The war had come to Odesa.

I figured this would finally make the news. I walked away from the party for a minute to call my piece-of-shit biological father, who I was still talking to back then, to let him know things were getting real. He started screaming at me through the phone to get the hell out of there. I told him I couldn't. It was where my family lived. I wasn't going anywhere without them.

There was no way that Natasha and I could go home that day. The building I had shopped at earlier that morning was on fire. Lena made room for us at her house and I thought about what the future looked like. We were in trouble and needed a new plan. Unfortunately, we couldn't save everyone. Even if we could, they wouldn't go.

When the morning finally came around, Natasha got us a taxi and we went home. I just wanted to go to bed, but it was her city and she wanted to walk around and see the damage. We had Volodia with us the day everything went down, so we knew he was alright. That said, we left him with Lena a bit longer to be on the safe side. Knowing that he was ok made us feel much better and we didn't know what to expect next.

America talked about how guns kill people, but it was complete bullshit. Not only could you still smell the bodies burning, but you could also see bricks torn out of the sidewalks and streets. The bricks were used to throw at people's heads. Even fucking 80-year old grandmothers had been chucking bricks. America didn't know shit.

By the time everything was over, more than 100 people were burned alive that day. Yes, I said burned alive, including pregnant women they raped and then burned. It was the greatest day of death in Ukraine since the end of WWII. Needless to say, we weren't really feeling great that day. I got some more champagne, but we didn't talk much.

You could tell from the way the bodies were burned, they just poured gas on their heads and lit them up. It was a horrible way to go if you asked me. You never know what man is really capable of until you see shit like that.

A little further down the road, they had barricaded up a building next to the train station and set it on fire. It was the Trade Unions building. The cops that had guns were shooting at people trying to jump from the burning building while people yelled, "Let them burn!" The whole city was chaos.

Since they were also looking to target Americans, we talked to our friends in Ukraine who offered to hide me in the catacombs. This was something I was not too keen on, but it was better than the alternative. Sometimes, things get extreme.

I had tried to get help even though I knew the effort would basically be feeble. The embassy's only reply was, "Don't try to come here because we can't help you. However, if you make it to Turkey, we might be able to do something." I sent an email to Senator McCain's office and didn't even get a "Fuck you." On social media, my former teammates and friends wrote, "I think they just Benghazi'd you." We were on our own.

Many of the structures in Ukraine were basically still from WWII. These guys were prepared. Additionally, many of the restaurants and bars were old bunkers underground, so I did a mental inventory.

Flowers were beginning to be spread out all over the city. The flowers were from mourners. Like them or not, the Ukrainians and Russians did have a bit of an ethical nature about them. This meant that with all the people that died, you took three days off to grieve and take care of the dead.

For three days, everyone could still smell the burnt bodies smoldering. I had lost friends that day. I didn't know where they went down or how they died. They were just gone. It's something that I am never going to forget.

Well, we tried to shake it off by focusing on Natasha's birthday. So, we went out for lunch. We chose one of the underground bunker restaurants that were close to our apartment. It was also close to the scene of the incident just a few days prior, but I felt better being in the bunker.

They did manage to take some prisoners during the massacre, but I think that was just for show. They had to look like they were doing something. Many more were in the hospital. I don't think anybody knew what happened to those people. I damn sure didn't. Supposedly over 200 were injured on top of the dead numbers. However, welcome to war.

We were having a good time eating pizza and drinking whatever, but I began to hear something on the street above us. I told everyone to stay where they were and went outside to see what was happening. Obviously, I wasn't sure how it was going to turn out, but it was better than one of the girls going out, or, at least, that was what I told myself.

The street was chaos. Although I didn't speak much Russian, it didn't take too long to figure out what was going on. The remaining Russians in the region had staged a backup operation. It was clear we were living among them and needed to be pretty damn careful.

On this particular day, I guess they decided to go back for the ones that didn't make it. I kind of respected them for their "No-man-left-behind" attitude. While we were celebrating Natasha's birthday, they were staging a prison break. They did get everyone out and the streets exploded with people yelling "Putin will never leave us behind." Shit was getting real.

With every passing day, it became more and more obvious that I was going to need a new plan. I already had a situation where I owned a house I never saw and a country that would not let my wife in. I also continued to remind myself I had people living in my house I didn't even know. Owning a hotel or bed and breakfast was not something I ever wanted to do.

We stayed at the restaurant long enough for the streets to clear out. When they did, Natasha's mom, sister, and niece made a break for it while the getting was good. They got home safely and called when they arrived so that made us feel better. We had far less area to cover in order to get home since it was just around the corner.

The next day, I began to feel out Natasha about leaving the country. I didn't want to say anything on her birthday because it was her day and she should have every chance to enjoy it. Volodia stayed with us that night and we got back into routine life, or, at least as much as we could.

The next day was a school day for Volodia. Since his mother and father didn't want to walk the streets, I took him to school. Basic services, like school, had begun to shut down. Things looked fairly normal when we arrived, but Volodia called me shortly after I left to go get him. He was one of only six kids who made it to school that day.

We were painfully aware of the dangerous situation that was growing, but I was looking at history in real-time. In that respect, it was kind of exciting, if you could forget about getting shot at and your friends getting killed. I guess you always have to look at the bright side of something.

Natasha continued to watch the media and complained as only a local would do. Support continued to grow around Petro Poroshenko who would become the next president. I just kept thinking about how hard his fucking job was going to be. It was also painfully obvious of all the bullshit in life that Natasha had gotten used to because she always acted like it was nothing new.

However, just because Odesa had heated up did not mean that focus had entirely shifted from the eastern region of Ukraine. The east experienced activity and deaths almost daily. Troops were occasionally captured and released after their weapons had been confiscated. One video did surface of a man being beaten and then buried alive, but, again, it was just a reminder of war.

I don't think that anybody felt that any good would come from it, but they did open an investigation into what happened in Odesa. All along, countries continued to have meetings and discuss peace. Russia continued to receive and issue sanctions and was eventually suspended from the G8 before they decided to leave the organization permanently.

Even though sanctions continued to get worse, it didn't seem to bother Putin at all. I couldn't help but think the world did not know who they were dealing with. As the Odesa investigation unfolded, it became increasingly clear that Russian nationals had been involved.

We were not shocked that they determined that what Natasha and I had seen was indeed identified as evidence of the staging of the event. Fuel and materials for Molotov cocktails were obvious items, but they even found chemicals like chloroform that had been left in staging areas. To nobody's surprise, firearms were also found to have been part of the staging equipment and many people in the police department were found to be involved in the assault against the city.

So, the month of May was progressing, but not in a good way. Both Ukraine and Russia claimed there were foreign operatives involved in all of the incidents. If you knew what to look for, you could see them on the streets and in the bars. In fact, the rumor on the streets was that with the election approaching, even the Ukrainian mafia was getting involved.

The election was to take place on May 25th. With all the anti-westerner sentiment, the Ukrainian mafia put out hits on all Americans. Putin maintained that he didn't have Russian troops in Ukraine. However, in a very clever strategy, he was catching Russian troops leaving service and offering them more money to become contractors and go into Ukraine. Of course, to sweeten the deal, those that said no because they had family suddenly didn't have that problem anymore and had nothing to lose. So, the USA had theirs, and Russia had them, too.

About this time, I contacted some professionals in South America to find out what we were going to need to get out of the country and get residency. Since I had lived in multiple countries and had become a Ukrainian resident, this was going to be harder than getting married. Natasha hadn't liked the documents then, and she certainly wasn't going to like completing more forms now. The list of documents was long, spanned a few different countries, and all needed to be translated into yet another language. We had to go back to our old buddies at the notary office.

Documents included things like our marriage certificate, our state and federal Ukrainian criminal record, and family information, including where everyone was born, copies of our passports and the family's, birth certificates, divorce records, as well as documents supporting what visa we would like to obtain for residency. This didn't even include the

American documents, but at least Natasha wouldn't have to deal with that. She got off easy, everything considered.

To make matters worse, some of these things were time-sensitive so those documents would have to be the last ones we got to make sure they would not be outdated when we filed them. The good news was that we had done part of this to get married in the first place. The bad news was Natasha was now in document hell again. None of this even counted the shit that got done wrong the first time or the other documents we had to "buy."

Keep in mind, we were in a civil war, with a deposed president, a new election, and government shutdowns. It was time for another beer. On top of all of this, the documents all had to be apostilled for international use. Luckily, our friends at the notary office had a "knack" for taking documents apart, changing them and making sure they looked "official" afterward.

None of this even included the problem of getting Volodia out or the documents needed for him as it was our intention to take him with us. This was a completely different issue. Unfortunately, since it was Natasha's son, she had to do all of that on her own.

Volodia was 14 at the time and wasn't supposed to have a passport until 16. However, Natasha and I pulled some strings and got him one. The unfortunate thing was Natasha let him take it to his dad's house and his grandmother quickly confiscated it. She ran the family and was basically like, “You are going nowhere unless we say so.”

We needed a notarized letter from his dad every time we took him out of the country, so moving was sure to be a pain in the ass. I loved Volodia and definitely wanted him to be my son. He reminded me of everything kind and innocent in the world, especially after the massacre. He also reminded me how the war was affecting the country every time he asked me how to fight.

It was surreal for me because many years earlier, while working for the USA under SECRET Clearance, something told me to have a bug-out plan. I began looking for other countries then. I wanted to really understand how people could change countries, what it cost, and how long it took. It was a weird moment, but a fact of life. At least, it was a fact of my life.

These are the kind of plans you have and never hope to use. However, if you do use them, you are always glad they were there. Natasha kind of looked at me like the RCA dog with his head all cocked sideways when I said I already had a plan for this. I mean, who walks around with an escape plan in their pocket? It was a bit different, but I have always been that guy.

I knew Russians that had moved to America and spent decades there. However, even they decided it just wasn’t a place to be after they became citizens. The so-called American dream turned out to be complete bullshit in many people's eyes. Oddly enough, I met most of them where Natasha and I would end up, which, of course, was Ecuador.

It was a complicated world that we lived in and I really wish I could explain the disgusting violence that took place in this situation. Ultimately, there is a strange soul-searching process that you go through when these events happen and your friends die. How you handle it and how you recover is a very personal process that you will never know until you are unfortunate enough to be in that situation. Maybe this book will help.

So, we embarked on our document collection process. We did start with the notary friends to see how much it was going to cost us. After currency conversions, it was several hundred dollars. It was kind of like putting together your illegal documents to flee the country. We spent many days with these people. They were already Natasha's friends, and became mine, too.

Trip after trip, the prices went up as we bribed more people for what we needed. Many times, as I mentioned, the docs weren't right. The official documents are sewn together and then the stitching was taped. So, our master document friends would have to pull the stitches out, fix the page, stitch it back up, tape it, and stamp it. A truly major pain in the ass, but the only way we were getting out of the country.

Every time we went to their offices, it was just a story of pain. They helped us every step of the way, but the bureaucracy was incredible. Luckily, they always had fucking cognac or I don't think we would have made it. Hell, one time, Nicoli even went to tweaking out my neck because I was so stressed out trying to get us to another country. We needed them and would need them again later.

As a side note on the cognac, Natasha had a high school reunion and went out one night. I didn't go. Eventually, I got tired of waiting for her and went out myself. With everything going on, this scared the fuck out of her when she got home to find I wasn't there.

We had changed apartments by then. It was a really nice place with a great bathroom and a deep tub. This was good because she spent most of her night in there that night. She called me and I quickly answered, explained I had only been gone about 30 minutes and would finish my beer and come home.

Well, Natasha, being nice, had brought home some cognac that night for us, but because she was freaked out about me walking around in a war zone, she downed half the pint in the ten minutes it took me to get home. It was nice that she cared, but she paid the price for it. I kept her head over the toilet for most of the rest of the night. I started counting every time she puked and I think I got up to 24. She hated her American husband at that point, but she was still cute and I loved her. Needless to say, she was useless the next day.

I couldn't really blame her. She was a beautiful girl with a new husband that had a target on his back in a country that had broken out into war, and we were trying to get out of the country. It was not an easy thing for her. Hell, just the reunion was an attempt at being normal since we had gotten married in secret.

It's kind of amazing to see people come together during times of war and destruction. I had seen it before, but each firsthand experience was always a bit different. On one hand, you have people trying to kill each other, and on the other, people become unified to help each other. If the world could ever figure out what really brings people together versus tearing them apart, they would be much further ahead. I personally think it is just character.

The election pretty much went as planned. There were no surprises and the man expected to clean house did exactly that. This, of course, led to more unrest. Politicians that were caught outside of their offices were thrown in trash dumpsters and rolled down hills. I personally considered it kind of funny, but I guess everything is relative.

Corruption was a big problem in Ukraine, being it was a new country. As I have probably already mentioned, they were never really independent until the fall of the USSR in 1991. There were going to be many things they had to learn for themselves, but I just kept thinking about what the early years of America must have looked like.

Getting out of the country was truly a three-prong approach. We not only had the Ukrainian people working on our stuff, but we also had the lawyer in Ecuador asking the embassies questions and showing them documents, as well as friends in America handling issues. This quickly became an international effort.

All along, I kept asking Natasha if this was something she really wanted to do. While the country was in a bad state, it was still her country, her home, and where her family lived. You can never underestimate the power of that. People will stay in a shitty situation their whole lives just because that is where they were born and grew up.

I told her I was willing to stay, but I think she was becoming increasingly worried about my safety and, of course, just wanted to see somewhere new. The more I told her we were going to the other side of the planet and she was not going to be able to go home every time Volodia sneezed, the more I could sense she wasn't ready. For me, it was easy to change countries because I had done it most of my life. Hell, I was already living in a new country. In the end, maybe I should have squashed the whole plan and stayed in Ukraine, but we will never know.

Back in Texas, there was always something that needed to be fixed at the house, even when I wasn't there, which was exactly why I never wanted a fucking hotel. I preferred businesses and careers based on intellectual property, and basically, knowing your shit than something that was inventory and capital based. Natasha and her sister had tried this type of business together in Odesa and were ultimately robbed and lost everything. Short of Jack Daniels and dementia, this would never happen with a service business. Ultimately, the day was getting closer where I would have to leave and go back to the states to take care of the things that needed to happen there. We weren't there quite yet, but we were already dreading it.

Chapter 9: President's First Days

Petro Poroshenko ushered in a new wave of hope for a failing, young, corrupt country in a civil war. For me, the idea of hope was hard to believe sometimes, but it never amazed me how many people counted on it. However, most seemed to think that one election was some kind of motherfucking magic wand, which was complete bullshit.

When he was elected, he was only the fifth president of Ukraine, three years older than me, and left a business that made him a billionaire. However, I knew he was taking on a thankless job. He was the first president after the last one had been deposed and run out of the country with many members of his team and family being killed. It was not easy being in this savior leader kind of position. I can't think of any time it has ever worked out.

That being said, people started counting the days from when he was sworn in. Just like the president of the United States, they expected him to change the world and end a war in 100 days or so. It was a very stupid proposition, but I saw countries do it all the time.

I had never been in any political office, but had taken on roles that I knew were fucked from the very beginning, so I knew this man had a tough job. I couldn't be sure, but I wanted to believe a guy like that only left his successful business because he wanted to help his country and cared for its people. Hell, President Trump didn't even take a salary.

I mean, let's be fair, the country was in serious debt and in the middle of a civil war. I am using the term "civil war" here because that was what many people considered it, even though Russia had already invaded the country. I'm not sure if it was the correct phrase to use for a combined civil war and invasion. Maybe it was just chaos and simply war. The IRA used to have a saying that was simply, "If you aren't confused, you don't know what's going on." Maybe that applied here.

I did know that many Russians and Ukrainians had escaped to places like Thailand and Israel and I am sure Poroshenko could have packed his bags at any time and bounced for the rest of his life. Instead of dealing with the bullshit of trying to be a savior president, he could have been living on a beach somewhere not giving a fuck at all. That would have been true whether his business collapsed, was run by someone else, or sold. In the end, the last thing I ever heard about it before leaving the country was that the business was going to shit and he tried to sell it, but couldn't.

That said, the new president did come out swinging. He knew he had to. I didn't know if he was ever officially sworn in, but if he was, it seemed like it took only ten minutes from the end of the election to get to work, unlike the months that you saw in America. Of course, that could have been because it was a special election and they needed a fucking president to blame for whatever the future might bring. In that respect, people are all the same everywhere in the world. They would rather complain about the leaders they had than stand up and lead.

So, the new president had the new president's playbook out and he was following it line for line. He knew what the country ultimately wanted, aside from the pro-Russia separatists. They wanted an end to the fucking war and corruption.

He ran his ass off all over the world meeting with heads of state and asking for help, but I couldn't see any major changes coming from any of them, even when he met with Putin. He spoke to the country and vowed to take back the east and Crimea, but what else was he supposed to say? In the end, this was, of course, easier said than done. As has been continually suggested here, Ukraine didn't have the resources or training to go up against Russia.

Ultimately, they ended up with nothing more than a stalemate, which was what I believed Putin really wanted. However, I thought that sanctions from Europe and America were still a pretty impotent response. It became the new status quo with lots and lots of talk, but no real progress.

The new Ukrainian president did make an attempt to attack corruption in the country and replaced a number of people, which always made the news, but nobody seemed impressed. Ceasefire moments continued to be negotiated and would last for a while, but after they continued to be ignored, with everyone pointing a finger at the other side, the ceasefire agreements became a joke. Maybe Natasha was right. Maybe Putin didn't really want Ukraine. It was no different than a hot ex-girlfriend; you don't want her, but you don't want to see her with anyone else.

However, the stagnation mentioned here did not mean that the president's first month was uneventful. He received a great deal of pressure with the war, and maybe, Putin was watching him to see if he would crack. I didn't know if Putin knew of Poroshenko before he became president, but I did know that Putin did not like to be caught off-guard and would test a person.

The unfortunate aspect of cock measuring between presidents, though, was that it was others that paid the price. With Ukraine, of course, it was not just military personnel that paid the price, but also civilians, as well, even if they got out of the country. The innocent died, had to leave, or lived in fear. Those who did leave had to say goodbye to family who may have very well died after their departure or they risked never being able to return to see them again.

So, the rundown on the first days went something like this. As said, the president vowed to take back the east during his early days. During this time, a commercial passenger plane on its way to Malaysia was shot down flying over the Eastern warzone, killing everyone. This showed the east was not under control. That got our attention! I was scheduled to go to Malaysia soon after that time. However, as one would expect, flights were immediately diverted away from that area of the world.

Again, there was a lot of finger-pointing. Ukraine said it was Russia and Russia said it was Ukraine. International teams showed up and the investigation went on for years. I don't ever think there was a clear conclusion or that anyone even cared by the time it was announced, except for family members of the dead and, of course, the dead.

Later that month (July 2014), a Ukrainian military transport plane was shot down. I can only assume that was what they thought they were shooting at before when the commercial plane went down. Some said these incidents were accidental mortar fire, but evidence showed otherwise.

In August 2014, the Russians entered Ukraine claiming to bring aid to Russians. This was always their excuse for invading another country and they were going to keep using it. By the end of that month, Poroshenko had ended talks with Russia and returned home to Ukraine.

In the end, at about this time, the bullshit was coming to an end for anybody in the world that was watching. As usual, this didn't turn out to be too many. I think that most governments counted on that when it came to their wars. Additionally, the rebels in the east began admitting that there were thousands of Russian soldiers fighting in Ukraine. In fact, ten Russian paratroopers were captured in Ukraine and basically said they ended up there because they were lost. I guess Russia's elite had fucking forgotten how to use a compass. Obama and others responded in their impotent style of sanctions. Afterward, another ceasefire ended, and emergency UN and NATO meetings were held.

This was ultimately how the war went while I was there, only without Americans protesting, because they basically knew nothing. It was no different than Southeast Asia or Africa. If it was out of sight, it was out of mind. The "You can't see it from my yard" attitude.

These are the times you become forced to face many hard decisions and find out exactly who you are. I had a beautiful lake house in Texas along with a boat and a nice car. My wife had been turned down for a visa to go to America and I had been out of the USA for over a year. On top of all of that, the war was now literally at our front door. Don't think it doesn't cross your mind to cut your losses and leave in these situations.

There was something else that drove me, though, and had all my life. That was what I had always listened to. I loved my house and everything that I had achieved. I didn't really want to lose it, but protecting my wife and family were much more important, and it was becoming increasingly clear that it wasn't going to happen from, or in, Texas. I could hear my grandmother saying, "It is only stuff."

This was something that my biological father never got or his sister. Maybe that is why my grandparents were ashamed of their first two children. They basically came out being people who only cared about themselves and stuff. At this point, my choice became obvious and there was only one thing I could do, even though I had kicked around the idea for years. However, just like executing every emergency plan, you always have some doubts.

Many years earlier, I had not only worked on building my actual dream home, but also had valences, curtains, bedspreads, nightstands, and other furniture sent from my grandmother's house to re-create a room in her house, so that she would feel at home in mine. We spent years there together and they were enjoyable. In fact, the room I recreated for her in my house had actually been mine when I lived with my grandparents. I went so far as to use the same wall plates for the light switches and the same paintings so that even when I was gone, she would feel at home.

It was a strange and frustrating feeling to have built something that you took a lot of pride in only to have to get rid of it and liquidate everything. I had to think about this for a long time. Natasha and Volodia were not going to be able to go to America for any reason at all, and I had people living in my house I didn't even know. The logistics of the situation were anything but easy, and this was only going to be the first step of getting us where we needed to go.

However, I continued to look at my wife and stepson while I thought about what the future held for us and had a beer or two. This was not a decision that happened in short order and was not going to be something I could reverse later. Needless to say, I thought about it for a long time. I am not sure if Natasha knew what I was thinking about at that time, but I guess even she knew something was going to have to be done. This was not a situation that could remain the same without major problems.

The truth is, a fucking war is expensive, and this war was taking its toll on the country and us. Job prospects in the country weren't exactly popping up for either of us. Natasha hadn't worked in over six years, and I hadn't been sent in by someone else, so we were surviving off of my own funds. I had a lot of skills and credentials, but still needed to figure out something.

I associated myself with every expat that I could find in the country to find out what they had been doing, how well it worked, and what their future plans were. Some were doing well, but most were just managing. The USA wasn't much better and there were even people flying to Ukraine to find out what they could do there. It seemed as if I was fucked either way and I really wasn't sure what to do. To say I was worried would be putting it lightly.

In the end, we only had a few options. The first was to stay in Ukraine, avoid the bullets and bombs, and figure it out, but I had a mortgage and bills and shit in America, so I didn't know how that would work. The second was for me to ditch the family and run, but I was not built that way and that wasn't happening. The third was to bet everything and roll the dice on what some could have called nothing more than a Hail Mary. I would need to sell everything to create cash, put together docs to enter a foreign country I had never been to before, and hope that I was actually smart enough to find some way to keep making money and build us a new life in a place where I didn't speak the language natively. Clearly, the odds were against us. I was feeling it, but we were going broke quick.

I had already been to many countries and seen many wars, but this situation was different. I would be risking everything including my life and the lives of my family, to get us out to another place none of us had ever seen on one-way tickets through a country trying to kill us. It was time for a beer.

I thought about the options for many days, but they all came back the same. I would have to take the chance and risk it all. I talked to Natasha about it and she agreed. The downside was, in a time and place where she wouldn't even walk her own son to school, I would have to leave them to get the plan kicked off. I had no idea how long I would be gone either. I started to plan my return to Texas.

That still did not fix the long-term money problem. We were going to need to do something for cash. We could get some short-term money for the sale of the house and other shit, but eventually that would run dry.

As I said, I had already been to many countries before, but there was always a plan and a check first. I never even paid for flights. For that matter, I hadn't even paid to get back to Ukraine from Saudi. We were now in unchartered waters working without a net.

I kept cruising the internet and asking everyone I could find what they were doing. Most of the responses were either oil and gas or teaching. The teaching thing did get my attention. I knew that many ex-government people did it in foreign countries when it was time to find a new place to live. Additionally, I knew teaching worked and with my credentials, including being a scuba instructor, it would be a start. It wouldn't be a lot of money, but teaching could be done online from any country in the world.

Of course, at this point, I was kicking myself for not spending more time on my backup plan instead of drinking beer during the prior years of having work. I had been working on a website for a long time to make sure I always had something going on. However, I failed to follow through to get it to the point where it could help me in times like these. I swore to never do that again.

I needed to do something, so I resumed working on my first book. It was a project management book. I wasn't exactly sure how I was going to use it, but I knew it could work with the website. The only problem was that it was not going to be immediate money, but instead, another expense. It was one more example of how things could not keep going on the way they were.

We've all heard the stories about the people that have made millions and even billions online as well as the people that had tried and never made a dollar. However, I kept hearing the same phrase in my head that said, "If one man can do it, another man can do it." It was the best worst plan that I could come up within a new world of shitty options, so I had to make it work.

I thought to myself how I had been writing code since I was 10 years old, I had two degrees in computers and electronics and one in business, and, on top of that, I had been recruited as a USA government contractor to work in electronic warfare when I was still in my early 20's. If I couldn't do it, then nobody could. I decided to try to find a way.

Additionally, I reminded myself of a time when I invested $2.50 in a Pac-Man book to learn the patterns at the age of 10 and ended up making hundreds of dollars while on the road with my father as he drove a truck. I would clean up in truck stops by betting on the game. I thought about how I learned to solve a Rubik's Cube in less than a minute. I knew if someone could put it in a book or on a website, I could do it. Just tell me how.

Well, at that point, the stage was set. I was soon to be off again. Natasha was kind of getting used to it, even if she didn't want to. We had a lawyer and started collecting the documents we needed. We, also, had a plan even if it wasn't the best plan in the world. It was the only plan we had, and sometimes, you just have to go with that.

Luckily for me, I had also been trained as an international project management expert. This made me adept at operation execution and had gotten me all over the world. Now, it was time for a real test, and we were all hoping that my training would pay off.

The rest of Natasha's family would be staying behind. We already knew that. Her mother knew what was going down. She wasn't very happy about it, but she wanted at least one of her daughters to make it out of the war zone. She constantly told Natasha it was obvious I loved her.

Of course, while all of this would be going on and I was out of the country, I would have to keep an eye on my newly acquired Ukrainian residency. It would only be good for one year at a time. I had to stay on top of it. As we were finding out, it is always good to be able to go to another country.

I didn't know what the real estate market was like in America or how well I would do with my house. However, this was the first project of our operation. I didn't even have a realtor, but I knew many and began reaching out to them.

My house was on the outskirts of Austin, TX in the hills near a small town known as Lago Vista. So, it was either going to work for people who wanted to be out of the city or not. As it turned out, the area happened to be growing. That was lucky for me, but I still didn't know how long the sale would take.

In some respects, I had even bigger problems with stuff inside the house. There were hand-carved tables, ivory and wood statues, monstrous mirrors, an Egyptian tapestry, paintings, and even a camel hair coat given to my grandfather by the King of Saudi Arabia during WWII. This was going to require something other than a yard sale. However, there were still many yard sale items also and I had never held one, so I needed help there also. This was going to require a whole team. Some of the furniture was so big it wasn't going to be moved without heavy equipment.

It was completely amazing that stuff I never hoped to lose became the largest pain in my ass I had ever known. This didn't even count things like the car, which I found out I could not take with me because of various laws. In the end, I was glad I didn't. I vowed to never own that much shit again.

Suddenly, I was jealous of the people that rented from me. They traveled light and only needed a room. They didn't need mountains of stuff or have to take care of a house and a yard, which was also going to have to be dealt with in order to sell the place. I made up my mind that after the sale of the house I would be a renter for the rest of my life. The rental income was helping, though. While I was in Ukraine, one room's rent in my house paid for our whole apartment.

On top of everything else, I realized that my US driver's license was about to expire at that time. It was not like I had been driving, but I didn't want to lose it. I was going to have to add that to all the other documents that I was going to need to take care of in America. What a fucking nightmare! Additionally, this was just the beginning of the operation. I began to wonder how I had ever managed to get around the fucking world with so many damn material possessions holding me back.

It was an amazing situation realizing I had spent years building something and protecting it only to make the decision to tear it down, part it out, and get rid of it. It was clear to me that my life would be forever changed, but I had no idea how much at the time. That was what adventure was all about, and I was embarking on another one with no intention of turning back. I was not a guy that made a decision on an operation and changed his mind. That just didn't happen.

So, again, Natasha and I stopped worrying about the moment and the decisions for the time being, and just spent as much time as we could together and with the family. I had grown to love all of Natasha's family and was worried about them, but I knew I couldn't save everyone.

This was not a really good feeling if you have ever had to go through it. I was getting ready to go back to what they considered a dream place and a paradise while I knew I was leaving them in hell. They couldn't help but wonder if I was really going to come back. How could they have not had doubts? In the end, we went to the airport and I assured Natasha and Volodia that I was coming back. I boarded the plane for Texas and got several drinks.

Chapter 10: The Last Trip Home

So, I had finally boarded a plane to go home for the last time. At this point, I don´t think I even knew where home was anymore and I probably still don´t. In this way, the old cowboy phrase, "Home is where you hang your hat" came to mind. I wondered what it was like to be that kind of man. In the modern age, I guess I had become that man and there was no going back.

It was an odd feeling going home under these circumstances. Some thought you should have died or even were dead. Nobody knew when you would come back or why, but everyone experienced some sort of surprise. It is almost as if you had been living on Mars for two years.

I, of course, had called a friend to pick me up at the airport. She was a good friend I had known for many years, and her husband had been involved in my first recruitment. Not to let the moment just fade away, we stopped at many bars on the way to my house.

She told me about everything I had missed and what was going on in her life. Of course, she asked about the war and I told her. She was happy I had made it back. I had to admit it did feel good to be on home turf again. However, I knew this time I was there to say goodbye.

It was truly mixed emotions. You are happy to be home, but you had hoped that this day would never come. Maybe that is why we stopped at so many bars or maybe I just wanted to see people after being gone for so long. In some ways, even if you are not captured, you are in prison. Coming home was freedom for a moment, but I knew it couldn't last.

Eventually, we got to my house and I said goodbye to my friend. I hadn´t told anyone else that I was coming home, which somehow had become the way I operated for many years. I don´t know if I didn´t think people cared, or I didn´t care, or I just liked to surprise everyone. I think it may have been the latter.

I walked into my house hoping that it would feel like home and started making my way to my room when someone emerged and asked, "Who the hell are you?" I said, "I own the place, so who the hell are you?" These were always interesting moments.

Ultimately, we did know each other, but only digitally. It is an odd thing how the internet builds relationships, but great all at the same time. We exchanged some small talk for a bit and then I moved my ass up the stairs to my bed.

I don´t care how many countries you have seen, when you finally get back to what you consider home and your bed you will sleep well. It felt like I slept for two days even though I knew it wasn´t that long. However, I eventually got up and found my buddy who had been watching the house and had a drink or three with him.

It didn´t take long before the magnitude of what I had to do started to set in. It also didn´t take long before the people that didn´t know I was back or wanted me back started to realize I was home. I quickly remembered why I didn´t miss the place until I got back.

I was in a completely shitty position. I had gone back to a place where I wasn´t wanted to take care of property there so that I could arm myself and go back to a place where I wasn´t wanted because I had married a Russian-born Ukrainian. Could I have been more stupid?

My family had acquired a great many beautiful things that had ended up in my house. This was apparently easy to do when you are a military family and owned an art gallery. I really loved the items and the history, but I was more prone to the military part of the family and specifically electronic warfare in my years.

I was learning what that meant even though my father never got it. It was treasure for him. Egyptian tapestries, gifts from kings, beautiful artwork, and frames painted in actual gold now had to go. On top of that, the home I thought I would die in would have to go. I needed to get rid of it all for the mission at hand.

I knew that had my grandparents been alive, they would have agreed to do anything to get Natasha out even though they would have known it meant risking my life. It made it easier for me to start selling and let go, but it was a lot of shit in a fucking 3-story house. I just got started.

I did achieve the acquisition of many things and it was funny how many people judged my success by my things. When I finally got rid of everything, I heard from many people how I had lost all my stuff. All I could do was basically ignore them and say that I hadn´t lost anything, but instead had sold it. Of course, many people showed up thinking they deserved something for free, like my biological father and his wife. It was always interesting to see these vultures that could never achieve anything on their own come to you with their hand out and later try to act superior in some way.

All that said, I was home with a big project ahead of me and I started planning. It wasn´t going to be quick and the more time it took, the more time they had to kill my family in Ukraine. The gravity of my situation started to set in.

I knew there would be many weeks of yard sales and this was never something I had participated in or organized, so that was the first place to start. A friend of mine in the neighborhood came to mind and I offered her 20% if she would put it all together. Luckily for me, she agreed. That was one thing checked off the list.

Next, I was going to need a realtor. Although I knew a few, the lake area was saturated with many who worked in real estate. I started meeting with people and showing them the house as well as figuring out what would have to be repaired or changed to get the hell out of the house. Obviously, I didn´t want to spend all my money on repairs since money was tight. I was going to need everything I could get my hands on for the exercise that lay ahead.

The house was a long way out from Austin. That was a disadvantage for anyone who worked there. On top of that, older people weren't going to like the stairs that came with a multi-story house. Lucky for me, many younger technology people were starting to move into the area who could work from home. Regardless, I had to price the house to sell as quickly as possible.

With the realtor situation taken care of, I now had to deal with all the artwork and furniture that needed to be sold, but would not be yard sale items. I spoke to some people and the idea of an estate sale came to mind. These people normally specialized in artwork and antiques and some of the furniture would need construction equipment to get it out of the house.

On top of all of this, I had a commercial grade phone system in the house, a home automation system with lighting control, and multi-room audio and A/V components for the home theater. My first reaction was to sell the equipment and leave the speakers and volume controls and shit, but in the end, the A/V equipment, some furniture, and even some artwork were included with the sale of the house. This just seemed easier than having a new homeowner trying to wire up their own shit. We did scrap the phone system, though. It was an expensive phone system, but the controller and phones just weren´t the kind of stuff that people used in homes. It was meant to run a business. It was painful, but there was no other option.

It took a whole team, which was hard to assemble, to complete the task, but it was one of the things I was good at. I had some hurdles, but slowly the team came together and there was hope. Hope was tough during those times, but definitely something I needed to keep going.

The estate sale person was the hardest to find. Art is a funny thing. You would hear in movies and shit what all this kind of stuff was worth, but it was only worth what people would pay for it. People weren't really into paintings and shit at that time. They were into photographs. The older people in America were scaling down to smaller places, so while they liked many pieces in the house, they had nowhere to put them.

The stuff had been in my family for a long time, so I didn´t just want to give it away. As I have mentioned, my biological father was anything but responsible. Besides, I didn´t feel like making his wife happy. Rather than lose some of the family heirlooms, I gave my father some items that were going to be difficult to sell. Other items were proving to be a bigger problem.

Living the way I had, you acquire some special items. This included mountain gear with water purification systems, throwing axes, specialty knives, and even gas masks. I also had a trauma kit that I kept around for injuries as well as a whole fucking lot of technical dive gear. I spent a lot of time on Craigslist and eBay.

All of this stuff didn´t even cover the flame-retardant Arctic gear that I didn´t really want to get rid of. That is the kind of stuff that you don´t need often, but when you need it, you need it. At that point, I had to start considering the storage and transportation of all my gear as well as how much I could scale down. This didn´t even include computers and other tech items such as video equipment, that I wasn´t going to leave without, because I needed all of it for finding work and making money.

Another problem I had to deal with was my office and library. Carrying books was not going to be the way to enter Ukraine, extract Natasha, and get to another country. There were even antique books, so what didn´t get sold got donated. By the time it was all said and done, I had gotten rid of most of my books and vowed to start using fucking electronic books even though I always liked physical books because that was how I had grown up.

In most operations that I have been involved in, during the decommissioning period, we had to deal with document destruction. This one was going to be no different. I had file cabinets full of shit and nobody was going to buy my old yearbooks or notes or letters or pictures or any other personal shit. All of it had to be destroyed.

It is one thing to do this for an operation when it doesn´t include personal items, but it is another thing entirely when all of it is memories of your life. Regardless, it all had to go and there was a lot of it. I started with the shredder, but it quickly became obvious that was never going to get the job done. I was going to need a big fire.

After a few beers and some thought, I remembered that there were BBQ grills and fire pits down by the lake, so I started loading stuff in the car to take down and burn. Lucky for me, the cops knew I was back and didn´t really want me back, so they didn´t get in the way. They knew I was there to clear out and leave.

Some of this gear was finally going to get put to a new use though. Obviously, Ukraine was now Russian-occupied territory and I wasn´t going to walk in there with nothing but my dick in my hand. I had already lost friends and had no idea how things were going to turn out when I returned. Additionally, taking any weapon across international lines could be a problem, but firearms would be a larger issue, so they were out. Furthermore, I couldn´t engage the whole country solo, so avoidance and stealth were key. This was going to need to be old-school.

I had a large waterproof footlocker-on-wheels that was meant for dive gear, but to me it looked perfect for re-kitting purposes. I had always been partial to edged weapons because they are silent, deadly, can be used up close and hard to trace. Given that I would most likely be dealing with improvised weapons in Ukraine, like Molotov cocktails, I needed to be quiet. This seemed perfect. I started loading up the box.

I didn´t pack anything that would be considered collectible type weapons because they could be traced or swords that would probably break laws for length, but any other edged weapon went into the box, including a crossbow. That was going to be enough to draw attention. I knew I was probably still looking at problems since my best plan was to walk right through the front door of a country engaged in a war with weapons, but it was the best bad idea I had. I figured I would just roll with it.

With all that in mind, I still had to deal with clothes. In America, we collect a lot of this shit. I must have had at least four leather jackets alone because I always wanted a backup, but this was fucking Texas! This didn´t even include all the blue jeans, business wear, and a shit load of ties. How did I get all this shit? It became painfully obvious that the world as we all knew it in America was nothing but material.

My bio dad´s dumb as-fuck wife was the perfect example of this. They were both in their 70´s and should have been retired, but she always wanted more shit. As if buying chairs and crock pots were like gold or something. The whole fucking place looked like nothing, but a junkyard, and yet somehow getting up at five in the morning to go to work, so she could buy more stuff to put in her trailer they lived in, somehow made her feel rich. They literally had 2.5 rooms in the place that were useless because they were stacked wall-to-wall and floor-to-ceiling with useless, worthless shit. She was counting the days until her mother died, so she and bio dad could go into more debt for a new house in their 70´s. However, that was America for you.

I wish I could say it was just these two idiots, but I would be wrong. My biological mother and her used car salesman of a husband, who called himself a tree doctor, did exactly the same damn thing. It was absolutely pathetic to see. Anybody wondering what happened to the kids that became useless adults in America should look at previous generations, like these morons.

I had been very lucky to grow up with my grandparents. They had more than all of them put together and never had junk rooms. I guess some people get it and some don´t. These are the things that I constantly thought about as I sold stuff off from my house and people showed up with their hand out looking for something for free.

Cash flow ultimately was the name of the game while I was selling stuff and getting ready to leave. Some of the money I lived off of, but I also needed to cut expenses. The obvious thing to me was to stop paying the mortgage on the house since most of the payment was just interest. Besides, I needed the money more than they did and was selling the house anyhow. It wasn´t like any of the countries I was going to go to were credit-based countries. Nobody had the debt of America. Hell,

Ukraine went to the IMF for less money than Detroit owed when they went bankrupt.

I wasn′t very skilled in foreclosures, though. That probably was something that I could have learned from my biological parents since both of them had filed for bankruptcy at least once. Remembering the aforementioned junk rooms, it was obvious why. So, while I was racing the clock, I didn′t know how much time I really had. I was inspired in many ways given this added situation. Shit had to get done as soon as possible.

I was in unfamiliar territory, but the definition of a project is "A unique effort with a finite beginning and a finite end" and I was highly trained in project management since it completely related to operational control. I was happy about that. It was going to take all my skills and experience to live through the situation. I had no idea how challenged I would be. All the training I had received over the years was about to be put to the test and I was only going to get one shot at success. To say I thought about it once or twice would definitely be an understatement.

Through all of this, Natasha and I had daily calls on Skype to stay in touch. I was glad that the internet hadn′t died in Ukraine, that way I could still know what was going on and talk to the person I was doing everything for. She was having some hardships though, so this meant that for me, I had to handle a war on two fronts.

With short-term goals laid out and being executed, I needed to start focusing on further out into the future. Some documents I could deal with at the time, but others I was not going to be able to handle until I knew the house was going to be sold and had a closing date. They were time-sensitive. This included being fingerprinted, again, and getting criminal records at the state and federal level, both of which would have to be apostilled also, so it was a two-step process.

Working to gain residency in a new country with all the documents that they required will convince you that once you are on the move, you may as well keep moving. It would be easier. However, there is freedom in a second passport and for decades I had my eyes on one from somewhere. I didn´t really trust America anymore after they left us in Ukraine without help. Having worked for the US government, I was one of the guys that had to learn the hard way.

For many years, it seemed like I was in a never-ending document process. First, I had to get documents to get married in Ukraine. Then, I had to get documents to be a resident of Ukraine. Finally, I had to get documents to get out of Ukraine to become a permanent resident of Ecuador. The digital nomad lifestyle started to look good. Maybe working online and keep moving was the best answer.

Based on how many months you could actually stay in a country with the passport you had, you could theoretically choose 2-4 countries to move between and keep making the rotation without needing residency.

Looking at the stack of documents I still needed, it didn´t seem like a bad idea. I even needed criminal records from Ukraine since I had become a resident there and got married there. I would not be able to hide that since we would be arriving from Ukraine. I hoped that my time in Saudi Arabia wouldn´t surface and become a problem, but I just decided to roll the dice on that. They basically wanted criminal records from every place I had lived in the last five years and I couldn't go retracing my steps. I had stamps and visas in my passport, but hoped they would be missed.

This, of course, meant it was time for another beer and some more thinking and strategizing. My beer drinking and strategizing time gave me the opportunity to start saying goodbye to people. Some of them also had some pretty good ideas, so there were always benefits to bouncing problems and ideas off other people. Many others were just happy I was leaving.

The house was listed with the hope it would move quickly without having to lower the price, but I was prepared to do so. Unfortunately for me, I had to drop the price about twice to get rid of it as quick as possible. I probably could have gotten the original asking price, but as I stated before, time was of the essence. We had no time to fuck around on this.

There wasn´t very much in the way of repairs I had to deal with, so that was good. I had been in the house for nearly 12 years, but it was well taken care of. That said, one dumb ass realtor showed up and told me I had to replace the toilets because they had toilet bowl rings. I was like, "Have you ever heard of a maid? Get the fuck out."

The carpets needed to be cleaned. Lucky for me, one of the guys renting from me had a buddy that did such things and got me a deal. The same renter was even able to help me with a cabinet repair. He was a good guy.

Things were beginning to come together, but slowly. At least, I could see progress. That was always a good thing.

One good thing about people that didn´t want you around was that they would do everything they could to help you leave. So, I had a network of helpers. I didn´t even have to pay them.

It was kind of funny, really. After I returned, the police departments would have daily meetings about me. It seemed they pretty much knew when I left the country and came back. One day during our yard sale, one of my favorite cops showed up and my friend started yelling at him to not arrest me jokingly. Something about she needed me for the yard sale. He looked at me and said, "I heard you were back in town." I was like, "Bad news still travels fast these days, huh?"

He was a good guy and knew why I was leaving. He didn´t know why I seemed to stay the center of attention for the police or what I did for a living, but he told me he believed I was on their side. America needed more people like him.

So, at this point, everyone knew I was back and why and the plan to get out was set in motion. The schedule wasn´t set and couldn´t be because there were too many unknown variables. I fell back into my old routine for a bit and kind of thought it was nice to spend the holidays in the states at least one more time. It had been almost two years since I had been able to do that. I missed being with Natasha during those days, but at least neither of us were alone. She had family with her and I had friends with me. The stage was set for things to get stupid as they often did.

Chapter 11: Still in Texas

At this point, I had barely made a dollar online, but I had started coding at the age of ten and had worked around IT my whole life. I knew it could be done. I had to prove it to myself, and frankly, I needed the fucking money. I decided to go for the low-hanging fruit.

Since I knew that one of the easiest ways to make cash was teaching English online, I started there. I knew that generating any kind of money weekly from the internet would put me in a whole new mindset. It was the kind of deal where you knew if you could make hundreds of dollars consistently, you could do a million.

In that way, it wasn't much different than learning how to walk. We had all heard the analogy before that you have to learn to walk before you can run, but it is true. Many didn't believe that anyone could make money on the internet because they had heard the stories of those who had tried before and failed. That scared the shit out of them. Fear is, of course, what keeps people from moving forward and breaking out of their chains, remaining prisoners of the system and their so-called careers or jobs. Freedom is what is beyond the horizon and, for many, always seems to be out of reach.

So, I started to cruise the internet and applied to many places and was finally accepted. I started teaching, and the money actually started coming in. Sure enough, I started thinking money could be made online. It was great! It wasn't a windfall, but it was regular, which was good. I was back to the concept that if you could make hundreds, then you could make millions.

At that moment, I was feeling pretty good. I had now entered the field of people who knew they could actually work online. That really got my juices flowing and thinking about the future. It also had me kicking my ass even more about the fact that I hadn't followed up on my website.

In my spare time, I began to review my website, considering where I would like to go with it or, at least, where I thought I could. Ultimately, since I had started teaching project management and it was a project management website and my first book was about project management, I figured teaching and consulting online about it was going to be the way to go. I could set it up as a monthly membership program. In this way, I would be in business anywhere in the world as long as I had a computer and an internet connection. It seemed perfect to me.

In between my work periods, I thought of Natasha, the future, and the war. It was going to be a long time before I got the smell of burning bodies out of my head and the quiet that fell over the city after the day of the massacre. I still tried to contact my friends wondering if they were ok, but never heard from anybody. It was clear that they had died. I had to accept that.

So, the planning and plotting continued while I worked the plan that was in motion to sell everything. I was making money, but definitely needed more. I mean I was still in America paying American prices for shit, so that meant a lot of money. I figured my internet teaching gig would probably be good in a foreign country, but in Texas, it just wasn't going to cut it. I couldn't lose it though, because it was an asset that I would need later.

In addition to everything else, I had managed to sell my bedroom furniture. I was now on a $40 Walmart air mattress, which was actually pretty comfortable, and I could just throw it away later when I left the country again. In general, the air mattress and I got along pretty well until the morning I woke up on the floor because it went flat. However, after a little work and a tube of superglue, I was back in business again.

Every step of the way, I was becoming more animalistic and angry. I knew there was a different battle coming. It wasn't much different than basic training. They took away your luxuries and gave you pain in return. I was ok letting go because I had come to realize all the things I thought were important were worthless given the job I had to do. Only time would tell in the long run.

Downtime is always the worst. I sat on the deck looking at the lake and thinking a lot. When I wasn't doing that or teaching, I was sleeping. I didn't really know if it was depression or if I was just tired from the last run, or both.

You always hear about the wars that stay with a man forever. This one was definitely going to be mine. It wasn't over yet, but I knew that this one would never go away. It was my fourth war, but this one was changing me. The problem was, given my moral responsibility, there was nothing I could do about it.

Every day I did the only thing I could do, which was stick to the plan and keep putting one foot in front of the other. I had started teaching in 2008 and I liked it. I guess I was finally at that age where I thought it was my time to pass on information and help others. Someone who had been very important in my life had done it for me. It was my turn to do it for others.

As it turned out, my mentor had died about four years earlier. I could have really used him at that time. He probably would have thought I was out of my mind even though he wouldn't have said it. No matter what, I never saw that man's blood pressure ever rise. He reminded me of my grandfather. Maybe that's why I had loved him so much. We would be out talking over problems and life and his wife would call to make sure he was ok. All he would say was, "I'm out drinking with Bill now, so I'll talk to you later."

INSANE IN UKRAINE

I had met Jim, the man that would become my life mentor, in the late ’90s when he worked for Tandem before Compaq bought them and they ultimately became Hewlett Packard. I was interviewing for a position with Jim and we hit it off immediately. Over the years, he taught me a lot and kept me out of trouble. He wasn’t a military guy, which was different for me, but he knew life and was pretty damn wise.

I had left the Arctic Circle and started teaching in places like Asia, Africa, and the Middle East. It was a pretty good gig, given the money I started out with when I was teaching in 2008, so, I went everywhere they would send me. This really helped convince me that teaching was the way to go and I started focusing more on learning how to build websites then.

I had begun thinking of a website in 2006, but got involved with starting a project with a buddy. Unfortunately, he fell on hard times when I was in Vienna in 2007 and died. After finishing the Vienna job, I started working on putting a site together, again. The year 2007, however, was tough for me too. It was the year my grandmother died. Basically, my web business idea had suffered spits and shits, so even though I was kicking myself for not getting it done sooner, I remembered the hurdles.

Training started taking off and I was making good money at $2500 a day plus expenses. Companies from all over the world started contacting me and I ended up in the Middle East again. I started getting calls from a number I didn't know. I ignored them because I was teaching or sleeping, but finally, I boarded a plane to fly back to Texas and my first stop was Atlanta. During my time changing planes in Atlanta, the number came through again. This time I answered. It was Jim's son. He told me that Jim had died and they hadn't wanted me to hear it from anyone else other than the family. I literally collapsed in the middle of the airport.

Jim had changed my life forever. While working on projects, his influence made me think more about passing on knowledge. Companies would send employees to my training seminars at $3800 per head plus travel expenses.

I videotaped all my classes and figured if companies would pay $3800 per head plus travel expenses for their people to listen to me, then $497 for a digital version of the course was a pretty good deal. I could sell it over and over again. On top of that, they would probably want copies of the slides, exercises, and maybe, even a transcript of the class. It was at least a plan to help us survive once we were out of Ukraine.

I thought about my grandfather and Jim and how they would like the fact that I never gave up. I would just keep working on the problem until I resolved it. They had trained me well. It was kind of sad that they never got to meet each other because that would have really been interesting.

There were still memories of Jim at my house. As I started to clean up the place, I found a bottle of scotch he had left behind. He knew I hated the shit so much that if it was the last thing to drink in the house I would either go dry or drink mouthwash. I never acquired a taste for it. To me, it always tasted like someone had stubbed out their cigarette in the glass.

My buddy and I had done some rough math on a class I did in Bangkok. It didn't take us long to figure out we needed to learn how to promote our classes as a company because there was a shit ton of money we were missing out on. We hadn't yet learned how to do that effectively, so I started pounding the digital pavement to try to find another class for some extra money. Things got pretty tight, but I was reminded of some old Rolling Stones lyrics; "...you don't always get what you want, but if you try sometimes, you'll get what you need."

When the fate of your event getting booked lies in someone else's hands, it pretty much fucking sucked. I mean, of course, they didn't make money unless they booked it out, but you always cared about it more than they did. After all, they had many events they were doing, but didn't understand the material enough to answer any potential student's questions. I had one place that was trying to promote events and didn't even tell the people about the speaker (me).

It was clear that face-to-face training was dying out because of the expense of travel and it was being replaced with internet-based training. Companies could save money, lower the costs of training, and keep people at their desks and working. It was as if only the truly privileged got to travel at that time. It wasn't over yet, though, so I was going to make what I could.

I did manage to get a couple of classes booked in October and November in Malaysia. I liked working in Malaysia because I didn't really have to go anywhere with the way the hotel and restaurants were laid out. They even had a speaker lounge, so if you didn't feel like running into students at the bar, you could always go to the speaker lounge and avoid them. It was also a good place to meet new people. Additionally, there was a cigar shop in the hotel and a Hard Rock Café right next door.

Most of the time, the food they supplied the class in the afternoon and lunch was all I really needed. The speaker lounge would usually have some small finger food kind of stuff, but I really didn't eat much. Malaysia was basically designed for shoppers.

The first time I was there, I walked around a bit to see the surroundings. It was mostly huge malls. When I say huge, I mean some of these things were like 10-stories high! I even saw one that had a floor that was nothing but jewelry stores selling Rolex watches. It was clear this place was completely designed for Arabs with way too much money. I went to a small store next to the hotel and got a case of beer for the fridge. I was normally jet-lagged anyhow.

Traveling for these things always presented an opportunity for surprises. I had a class in Bangkok once and my buddy came with me because he had never been there. That one started out a complete mess because my itinerary was all jacked up. I missed my flight to Japan and was re-routed through France. Instead of getting there at about 9 p.m. the night before I was to teach, I arrived at about 6 a.m. the morning of the class. As if that wasn't bad enough, my bags had been sent to Italy. At least I had my computer bag with me so I could teach the class. My buddy had arrived the night before and loaned me one of his shirts. My bags showed up that night.

We went out after class, but didn't go far. After taking a left immediately outside the hotel and walking about a half block, we came to a road with a shitload of bars on it. It turned out to be the very street where they supposedly had filmed the Bangkok riot scene in Hangover 2. We always seemed to have a knack for finding places like that.

On this particular trip, my buddy asked me how I traveled all over the world and never got sick. I told him to always drink beer and eat only fried food. He didn't listen, but he wasn't the only one who got sick on that trip. Some of the people that got sick were from Southeast Asia.

On the last day of class, we caught another plane from Bangkok to Phuket where we stayed for four days mostly drinking and partying. There was one exception. We went diving one day out of the four near the Burmese border. It was good diving and we met some new people that were all cool. Everybody seemed to be expats between jobs, so I was beginning to come to the realization that people like me weren't meant to have a steady home. One guy was even Canadian who mentioned that he was going to have to explain to his wife how he got from China to Thailand. I was glad I wasn't him.

We flew back to Bangkok the night before our stateside flights were supposed to leave. Both flights were pretty early in the morning. At the airport, I became friends with someone I had just met and who turned out to know Ecuador very well. He even had land there. I got all the information I could out of him about the country. He felt I was on the right track with what was still just my backup plan at the time, which was good news. On that particular trip, because of the Japan fuck-up, I ended up going around the world in ten days. To lessen the pain, the crew gave me a free bottle of champagne even though the drinks were free.

The Malaysia trip in October, however, turned out to be pretty uneventful. Uneventful can be nice, too, because crazy can be, well, just crazy. The taxi rides back and forth between the airport and hotel were always a pain in the ass though because it took about 45 minutes. However, I didn't fucking care since I was not driving and the hotel was cool. They even left a cake in my room once for my birthday. I remember that trip because I was still talking to my dumbass biological mother who asked me what the 4th of July celebration was like in Malaysia and I had to remind her that was America's Independence Day.

I wrapped up the class in Malaysia and started playing with video testimonials I could use for my website. Once again, I was on a plane headed back to Texas to see how things had come along while I was gone. Everybody involved in the project for me to leave the USA were people I could trust, so I figured everything was going pretty well and that turned out to be the case. That didn't mean that some of them didn't feel like killing each other from time to time, but that is just the nature of teams, so I knew it would be ok.

On the November trip, I wasn't as lucky when it came to uneventful trips. Getting there and being there was ok, but then things got a bit stupid. This was often a problem when you walked around in a perpetual state of jet lag.

I don't really recall what I was teaching that trip, but it was something project management related and it was another 3-day class. Most of the students, as always, were from around Southeast Asia or the Middle East. I was beginning to understand the culture and that area of the world very well.

This time, flying from Austin, I was going through China both there and back. I didn't have a visa to enter China and go sightseeing, but their airport did have a transit zone that was visa-free. This, at least, made the trip more enjoyable.

During this trip, I discovered the lounges, which were very nice. I had some pretty big layover times and hanging out in the lounges beat sitting in the terminals. I thought the club lounges were just for premium fliers, but it looked like things were changing.

For about $50 you could get a 1-day membership, which included food and drinks. On top of that, they had these great lounge chairs to get a nap in and free internet. Since $50 in drinks and food would have gone pretty easily, I was in.

On my way back, my bag started falling apart. I had gone through about three of the damn things because I really liked the design, but on this trip, the handle flew off the thing. Unfortunately, it was my main carry-on bag that held my computer, so it kind of made life a pain in the ass. However, it had lasted a pretty long time like the rest of them.

After some surfing the internet to stay caught up on life and Natasha, I got what I figured was at least $50 in food and booze and caught a nap. When I awoke, I was not only still half asleep, but also jet-lagged and made my way to the plane for my next leg of the trip. The next stop was Houston, TX.

On the plane, I engaged in my typical activities, which were usually watching movies and having drinks. Sometimes, I actually got to see things I hadn't seen before, but most of the time, I was watching things I had seen many times. While I was tired and needed sleep, I normally didn't sleep on the plane when I headed back to the states. It wasn't because I couldn't sleep. It was more about the timing of it all with the jet lag.

I have no idea when we finally rolled into Houston. After a while, it just all became one big blur. I do remember I didn't have long to get through customs to make my connection to Austin, so I was running my ass off with a carry-on bag that had wheels, but a broke handle.

I finally got to the magic fucking metal detector and had to break down my bag and take my computer out. I went through the routine like I had done all my life. Only this time, I was still hanging on to the fucking handle for the bag. I guess I was still thinking about fixing it later.

Well, as luck would have it, sure as shit, I ran off to the next plane and ended up leaving my computer at the security checkpoint. I got home and passed out and didn't notice it missing until the next day. The computer was LoJacked, so I was able to figure out where it got left.

I called the airport in Houston and told them exactly where it had been left. There were receipts in the bag with my name on them, which would identify me. I called for several weeks and even filed a report, but the answer was always the same and they said they didn't have the computer.

I needed a computer so I could work and I also needed to get a notebook computer for Natasha. I broke down and bought another one. It was actually pretty nice and I started to like it a bit more than mine, but I did still have a shitload of data on mine I wanted back.

I got back to work on the new computer and started making it my own with all the personal settings and software that I used. Sure as fuck though, the damn computer that was missing was right where I said it was. About 4 months later, the airport contacted me and they had found it. It was right where I had told them it was. They were able to identify it because of receipts they found in the bag, which I had told them were there. You just have to love it when people demonstrate how good they are at their jobs. At least, they sent it to me for free.

Basically, we continued working on the plan. The two classes in Malaysia had given me some money to ride out the storm until the house sold. However, for many reasons, being in a house that wasn't liquidated yet continued to create stress for me, especially since my family was in the middle of a fucking war zone. The only way to get past that stress was to fix the problem. That was all I thought about when I went to bed every night. I continued to remind the team what was at stake every minute I was in the USA.

Chapter 12: Wrapping Up Texas

In Texas, the plan was beginning to come together. Shit was getting sold, repairs were being identified and reconciled, and open houses had begun. We decided it was the deck and the view that was going to sell the house and it just may have.

As I said before, the house was priced to sell because I needed to get the fuck out of dodge. We luckily listed it at the right time of year since people with kids didn't like to move once school had already started. We were hoping that would help us. As I mentioned, had I been able to hang out and wait for my price, I would have gotten more for the house, but I didn't have the time, so we would periodically drop the price to be aggressive.

Slowly, but surely, I began saying my goodbyes and I spent time with everyone I could that meant something to me. I knew that once I left, there was no telling when or if I would ever go back. There were people in other states I couldn't get to, but that was just going to have to be a cost I had to pay.

I kept staring at the large furniture in the house, though. I hated selling it with the house, but with the way people that lived in America had changed, I just didn't see any way around it. The house we had in Florida, where the furniture came from, wasn't really that big, but everyone kept saying although they loved the stuff, they didn't have room for it. The large furniture became permanent fixtures at my house.

In the end, the furniture was one less thing to deal with. I still had a heavy-as-hell mirror to get rid of and a Persian rug that had dents in it from my bed. Who knew that ice and a toothbrush would fix shit like that? The internet turned out to be useful for info like that. The whole time, I kept asking myself how I ended up with so much shit. I know I have already said that, but I asked myself more than once, also. However, bio dad and his dumbass wife felt stupidly rich with every little trinket they got.

The bio dad and I did finally sit down for a beer after I returned. I told him I believed that he didn't think I would make it out. He said that he believed I would, but he didn't know how many holes I would have in me when I got home. Either way, he couldn't really believe I was going back. None of it changed the job I had to do and I got back to work.

That was always one of the things about me. I figured out a long time ago the only way to accomplish anything was to make up your mind to do it and then just do it. That's it. There's no big fucking mystery. The people who tell you of all their challenges just didn't want it bad enough. This was why the world had always loved stories of those that did it. Even in WWII, the mentality of the troops was like, "We are not coming back until it is over." By the 1990's, soldiers were killing themselves because of a 6-month tour. People got weak.

That may offend some people and if it does, so be it. The truth was, most of the world was laughing at Americans every time they saw them in movies and shit on a couch, talking to a shrink about their problems, and paying for it. The rest of the world was like buy a beer, suck it up, and maybe talk to a buddy, but we still have the same fight tomorrow. So, at a time when America was suffering from an overwhelming amount of unjustified privilege, I was getting ready to go back into a place where I was not wanted, doing it solo, and facing imprisonment, torture, and death.

I avoided thinking about all that bullshit by loading up on alcohol and working the plan. The sick thing was I had to really work to get there. I was in the process of selling everything in a world that thought I had gone crazy. I wondered what other people would give everything for or if there was anything they would give everything for. You never know what you will do or can do until you are tested.

Activity on the house began to heat up and we were really starting to get hopeful of a sale soon. I continued to tell Natasha that it wouldn't be long and planned my return to Ukraine. I was extremely apprehensive about going into the country again with weapons, but I just couldn't see going in without them. I had to do something.

Naturally, realtors don't want the owner of the house in the place when they are showing it, and we were doing open houses basically every weekend. My realtor was great and tapped a number of people in the area he worked with to help move the place. I respected him for that. He was former military and had a lot of honor. He was just a good man.

The yard sales and repairs were done, so the real estate team rolled in and ran the open houses while I escaped to the golf course. Everybody there knew me, there was internet, and, of course, booze. I continued to grow my internet training experience and I have to say that it continued to be a nice feeling to know that you could make money online. They damn sure got used to seeing me at the golf course, even if I was asleep in the sun, and I started liking the teaching thing. I guess I was the right age for it by then. I liked my students and learning about them since they were from all over the world.

I worked with a lot of Saudi's, at first, since I had just been there. The new King said that if Saudi's wanted a job or a promotion, they would have to learn English. That helped me with my efforts, but you would be surprised how many people would contact you thinking they could learn English in a weekend or something. Fuck, I had spoken English my whole life and still had problems with it.

While I spent most of my time turning the golf course into my private bar as the realtors showed the house, I somehow ended up showing the couple around that eventually bought the house. I can't remember why I did. I think it was because they were asking about all of the electronics in the house or they just showed up before I left. It was just the wife at first, but once she got to the deck and saw the view, she was on the phone with her husband who showed up pretty damn quick.

My realtor was there, so we loaded up in his Jeep and showed them the rest of the area as I greeted people along the journey. They must have liked what they saw, because before we ever stopped showing them around, the husband used his phone to send in an offer for the full asking price. We were over a major hurdle.

Of course, the house had to be inspected like parents looking at their virgin daughter on prom night, and sure as fuck, they found something else that had to be fixed. It didn't cost much to repair, but I was broke. The realtor, being the great guy he was, said he would cover it and would get the money back at closing. We were on. The deal was going to get closed.

I called Natasha to let her know we were really close. I was also preparing for a war I knew I couldn't avoid. It's an odd thing being trained to fight, but knowing you never want to have to use those skills because of the results. It is never good when you have to be prepared to kill someone's father, brother, son, or friend. I started to get my mind in the right place, which, of course, required alcohol.

We were on the short-timer list now and could see the end zone. We just had to make the closing happen and get close out of the operation completed. I started to ditch all the last stuff like a horny, drunk girl on

Friday night with a dress on and no underwear. It was flying out the doors.

Given the challenges I had to overcome, this was a good spot to be in. When I first approached the estate sales guy, he didn't even want the gig because it was too far out of the city. When I realized he was absolutely my best choice and the only option, I contacted him again and explained the Ukraine family problem. He jumped on the bandwagon like a true Texan and decided to help make it happen.

Unfortunately, the war wasn't nice enough to wait for me. While all of this was going on in the states, things were getting worse in Ukraine. The rebels had gone after the coastal city of Mariupol. Many thought it was a move to create a land bridge to Crimea and take the entire east. It ultimately failed. Had they succeeded with their plan it could have very well included Odesa, which would have made my job much harder.

There was a ceasefire agreement in January of that year (2015) and another in February, but it didn't really seem to make sense because I don't think it ever got to the ground forces since they kept shooting. They obviously didn't care. Later in February, Odesa was directly targeted. I was looking at a real problem since I was flying into a fucking airport that only had two gates and toting a footlocker of edged weapons. My sphincter factor was up to about a 10.5.

The fuckers in Donetsk, who were pissed about their city, were threatening Odesa on video. They were pissed, and probably had a reason to be, but they had started the fucking war. The airport there apparently was really nice and had held a special place in the hearts of the people of Donetsk and they wanted to return the favor. Odesa, where Natasha and I lived, was basically considered the crown jewel of Ukraine, so in the video, the rebels said they wouldn't stop until Odesa was destroyed and the pro-Ukraine people didn't deserve a truce. While I was waiting for my home sale to close, hell was headed for my door.

While everyone around me knew I was stressed and why, I never really gave them details about how serious it was getting. That was my problem to deal with and not theirs. At least, that was how I felt about it. It seemed to be a trait that ran in the family, so I guess I was trained that way.

My grandfather had personally met and worked with historical figures and presidents and became friends with them. Whenever I asked what he did, he said he cooked. The man went from private nobody to a major in 15 years. I would say, "That must have been one hell of a pot of spaghetti Major," as if to say, "You lying sack of shit." However, that was our way in the family. You just got home and they asked if you had any wounds that needed to be treated. Then, we went on with life.

People who didn't grow up this way would never really know what it was like to be in a military family. You had to get tough or die. My aunt became a very weak person and hated the family. She was crushed under the pressure, I guess, and never became anything but a bitter person who hated everything and everybody and had no personality. However, it was not my problem.

April of that year was a big problem. We had ended up with a contract on the house, but things were still rocky. The USA decided to get a bit more involved with what was going on in Ukraine and decided to send in paratroopers while Russia warned of escalations. On top of all of that bullshit, we had storms out the ass in Texas that took out the roof of the house that now had a contract on it and also caused deck damage. I was waiting for motherfucking locusts and the plague next, because it seemed to be problem after problem. In fact, even the fucking bar I used to go to burned down at that time.

So, while I worked on the next round of house repairs, looked for a new bar, and dealt with insurance companies, I continued to marvel at the feeble response to the Ukrainian war under President Obama and the EU. It was like they had all forgotten history and commitments made. However, this was politics.

It was clear that Putin had absolutely no respect for or fear of Obama. Why should he, given the response that Obama delivered. It made Americans, as a whole, look weak and may have actually cost lives. I personally think Obama didn't give a shit about Ukraine or America. It would have been interesting to see the situation had someone like Ronald Reagan been president at the time.

Political stuff aside, I tried to keep focused on the plan that I could control. In the end, it was all I could really do. America was not under my control and while I had given interviews in newspapers, that was not going to change the world. It was quite simply a time to embrace the suck and just keep going.

In fact, in one of the interviews I did, I mentioned that Russia was always strong when there was a weak president in the White House. I had said that Putin was a strong leader, unlike Obama, and that, of course, led to people in America calling me a "communist." However, many people were like that. They had no idea that I could consider someone a strong leader and still not like what they were doing. Then again, less than one-half of one percent of Americans had ever lived in another country and barely 1 in 3 even had passports and often did nothing with them.

You meet and experience many people when you are the traveling kind, so I tended to take them all in, even though I only embraced some. I guess that had put me on a pretty short list. Oh well, I had gotten used to it. I was always on someone's short list or shit list depending on what they actually thought of me. Sometimes, I was on both lists for the same person.

When you are trying to inform people about a situation you know they are unfamiliar with, you can't be afraid of telling the truth as you see it. I was the one there and experiencing the situation and they were not. There was a possibility that they would not see things my way because it shattered their view of the world, but they were not there going through it with me, so fuck them. This was my journey and they could either learn from it or not.

So many people in this world got their information from government-controlled news and then told everyone that they knew what was going on in the world. In reality, they just got information that was processed for them to make them think a specific way. Maybe they knew this and maybe they didn't, but either way, they still seemed to have blind faith in it. I presumed this was part of human nature.

This became obvious to me when I would talk to people back home, some of which were of Ukrainian descent, whom I had known since high school. They were truly good friends and brothers. They knew I was there and were concerned about me, but they were quick to tell me every time we talked that they were happy that I posted news about the country. Up until the Odesa massacre they hadn't really heard anything. Their statements provided further proof that the information was being filtered and distributed as someone else saw fit to do it.

I had to learn to have thick skin for many reasons. The men around me were tough, the enemy was tough, the scene was tough, losing friends was tough, and dealing with people who didn't know or understand was tough. In this way, it wasn't much different than project management.

Don't get me wrong, having a thick skin didn't mean I didn't care. It just meant I was going to do the job regardless, had learned to pick myself up, and I would stick with a decision and own the results of that decision. I had gotten in trouble many times in my life, but my mother (my grandmother) always told me to go in with my head high and stand straight. When the government recruited me, she just said she understood why and to be careful. According to her, I wasn't much for following rules and didn't mind risks and she could see where that would be useful. She was always the perfect officer's wife.

As if all of the things that had already happened during this time were not enough to drive someone crazy, there was still more shit to come. It truly was getting pretty rough, but there was nothing to do, but keep going. It was the only way I was going to be successful with the endeavor.

In April of 1986, nearly the very same hour the Chernobyl incident took place, I was heading to the hospital. I had been involved in a major accident that left me learning how to walk again for six months or more and changed my life forever. Looking back, it was almost like some bad butterfly-effect movie, since Chernobyl was in Ukraine. The country and

I seemed to be inexplicably intertwined. Of course, right now, you are asking yourself, "What in the fuck does that have to do with this story?"

Well, Chernobyl was still hot and forestation had continued to thrive, for some reason. Yet, these trees were full of radiation. In April of this particular year (2015), nearly 30 years to the very day, there was now a forest fire spreading through the area, burning all those fucking trees and releasing the radiation into the air. As if my job wasn't already hard enough, I was now going to be going into a territory with radiated ash from a motherfucking Chernobyl wildfire! I simply had another beer, laughed, and shook my head while staring at the lake.

If this was a tale of fiction, I could have written motherfucking Harry Potter, but I just can't make this shit up. The hits just kept coming and coming. I gave up trying to think about what could happen next. I was damn near afraid that some cosmic force could read my mind and come up with something even worse than what I was thinking.

I went back to working on the house and getting ready to get the hell out of the USA. There wasn't much else I could really do. At that point, things were too far along to turn back and my family was still in Ukraine. I just drank and tried to stay busy enough to keep my mind off of everything.

By that time, and really all along the way, I was now not only saying goodbye to friends, but also my possessions and a way of life that I thought I would have until I died. I knew I wasn't the first person to ever be in that position. I was alive, which beat the alternative, but it was a lot of changes in a short amount of time. Reflecting on my life, it amazed me how much it had changed in less than two years. Equally amazing to me, was how the Ukrainians had gotten used to a life in constant turmoil, which would have wrecked the average American.

As I cleared out the house by packing items or selling them off, every step of the way I remembered where the items had originally come from, which included several relationships in my life that were in the past and never would be a factor in my life again. Maybe this was why some people didn't ever want to sell their houses and things. Despite being inanimate objects, they brought back memories, dredged up emotions of those they had said goodbye to or reminded them of times in their past they didn't want to think about. I remembered how my grandmother never buried my grandfather's ashes or ever really got rid of his things when he died. That didn't happen until she passed away many years later.

That memory reminded me of my bio dad's multi-day drunk that he went on after having driven from Texas to Florida because his mother was dying. He showed up about two minutes too late. It didn't really matter though. I was the last one to have talked to her and that was in the middle of the night after I flew in.

She asked what I was doing there. Grandmother knew I had been in Austria and other places. I jokingly replied that I didn't have anything better to do in the middle of the night. We talked for about an hour or more. She was still pretty alert, but tired.

When I left to get some sleep, I told her I would be back in the morning. By the time I got there, however, she was unconscious and death was at the door. They handed me a Do Not Resuscitate (DNR) order to sign. It was hard to do since I was never trained to quit or give up on someone else, but she was 93 and had enjoyed a long life and I knew it was what she wanted. I stayed until she passed and then I moved on. With the sale of everything that had linked me to them, that life was definitely going to be behind me.

I thought once again about how none of us knew what tomorrow would bring. How I just needed to roll the dice again and go for it, as I had always done. I made up my mind that if it came down to it, I would die fighting before I would go to some Russian or Ukrainian prison. I already knew that America would not be coming for me. They had already made that painfully clear. I packed up that corner of my mind and put it away forever to never be accessed again.

Chapter 13: Back in the Shit

Well, at that moment, the Texas project was complete, or at least it was until something changed. I had put shit in storage until if and when I got out of Ukraine again. The only thing left to do was to finish packing up my shit, including my footlocker of edged weapons to take into Ukraine, set a closing date for the house, and make sure all of the paperwork needed to enter Ecuador was in order. Of course, I was not going to leave any beer behind, so I needed to make sure that was polished off, too.

The day of the house closing finally came and I had scheduled my flight for the following day. Since I had sold damn near everything, including my car, the realtor was nice enough to get me to the closing so we could all settle-up and sign a shit ton of documents. Afterward, the realtor dropped me off at an airport hotel I had booked where I could get a shuttle to my flight the next morning. After having a beer or two and trading some stories, he wished me luck and left. I just finished the night by drinking more after checking into my room.

It was the first week of May and the closing happened on 5/7/2015, so I had done pretty well with the schedule given my original estimates. However, there was still one stop to make on my way back to Ukraine. I needed to pick up some more gear and see a few more people before I left. My first flight was to my old hometown of St. Petersburg, Florida.

Florida was a pretty short hop, about 3.5 hours, from Texas compared to the rest of the trip. I wasn't going to stay long, so I began booking my flight to Ukraine. This was a one-way flight. In many countries, if you entered with just a one-way ticket, they would deport you back to where you came from. Obviously, I was hoping it was going to work out. After all, I was a Ukrainian resident. Now, it was just time to deal with the Florida business.

I hit Walmart, the western store, and the military surplus stores to get what I needed and still had room for in my bag. Then it was time to catch up with some people and say a few more goodbyes. I hadn't really returned to Florida very much since I had left in 1989. It had been eight years since my last trip back, so there were many people on my list to catch up with. As timing would have it, one of the guys on my list had a birthday coming up while I was there, so he was the first one I visited.

I had known the birthday boy since high school. We were good friends that always stayed in touch wherever I was. While I hadn't been in Florida for eight years, I had seen him in Texas when he came to visit, so it had been a while since we had seen each other, but not as long as most people. As always, when I got there, we picked up life right where we had left off. We both asked each other what the fuck had been going on, what was new, and in the process drank a shitload for his birthday and got pretty fucked up. He, of course, wanted to know how I ended up in the places I had and why the fuck I was going back.

The next place I had to stop were also friends that I had known since high school. They were American, but of Ukrainian descent. They had followed the situation more closely than most, knew I was there, and were worried. It had been much longer since I had seen them.

The brothers were happy to see me, glad that I was ok, but worried about me going back. They also had a lot of questions. We spent most of the night talking and drinking beer. After finding out I hadn't made arrangements to stay somewhere, they asked me to stay, which I did.

While going from house to house and place to place, people quickly found out I was in town for a while. As it turned out, a buddy I knew from middle school, but hadn't seen in a very long while, was also in town. It didn't take too long for someone to get the word out about a larger event for all of us to get together since two long-time out-of-towners were back at the same time. It was a nice event and great to see everyone and things continued like this for a few days until it was time to go.

Most everyone knew I had been in Ukraine, so once again, they all had questions about what I would do next. I gave them the rough plan. Concerned, they wanted me to just stay home, but they all understood why I couldn't. Knowing me as they did, they knew I would never leave Natasha in Ukraine. I just wasn't built like that.

The next morning, I was so tired and hungover, actually, I was probably still drunk, I almost overslept and missed my flight to Ukraine. I started down the road for a minute and realized that I hadn't even packed up all my clothes. I had even left stuff in the closet! I had to turn around and go back. I was so out of it that I left the key to the rental car in my pocket when I returned it. However, the airline was nice enough to take it back for me.

As usual, it was going to be a long flight and I would have to go through Turkey. I hated the airport in Turkey, for the most part, because the internet sucked even though the bar was good. Sometimes, I would even have to get a room at the airport hotel and stay overnight, but at least I was headed back. I settled into my seat, got a drink, and went to sleep.

I had managed to get everything in Florida done in about two weeks and headed for Ukraine on 5/23/2015. This time, I worked on scheduling everything so that I wouldn't have to stay at the airport hotel in Turkey and would arrive in Ukraine during daylight hours. I still wasn't sure how going through customs and the X-raying of a footlocker full of edged weapons would work out, but I figured that I had a 50/50 chance. Odesa was pretty much split straight down the middle when it came to pro-Russia and pro-Ukraine supporters, which, of course, made life very interesting.

Natasha was set to meet me at the airport. We were just going to have to find out how things were going to go down. Since she was a former lawyer for the Ukrainian Ministry of Interior, it would have been nice to have her with me during that part instead of me struggling to explain the situation in Russian, but I didn't have any such luck. I was going to have to run the gauntlet and get through security before I even got to see Natasha.

Sure as fuck, all of security came unglued when they scanned the footlocker. I gave them my Ukrainian documents, explained that I was a resident, and had gone to the US to sell off my things to move to Odesa. After some real pucker time where I didn't know how it was going to go down, the senior man finally showed up and said, "Fuck it, he's Ukrainian." He pointed out to me that I may have trouble getting out of the country especially if I was still carrying the same shit. However, at that point, I was just taking things one step at a time. Getting out was going to be something to consider later since this was a fluid situation.

I was sweating my ass off from humping all my gear since the 2-gate fucking airport didn't even have a luggage cart and found Natasha right outside the door. While she watched the shit that I brought out, I made a few trips back in for the rest. I kissed my beautiful wife and asked her where the fuck we lived now. She had moved us again. She had learned a moving target was harder to hit.

In typical Natasha fashion, she had us city center within walking distance of all the cool spots. Also, in typical Natasha fashion, she had gotten one of the best apartments and views in the building, which meant the top floor. In most cities, I wouldn't have minded, but once again, this was another fucking building with no elevator and we were on the fifth floor. After a couple of runs up the stairs, a guy came out of his apartment to find out what all the noise was about. Natasha got him to help us. We got in, got settled, and got naked.

For almost a week, we didn't do anything to prepare to get out. We were just happy to be back together and wanted to enjoy the moment for a while. After everything it took to get that far, we thought it was time we deserved. I was jetlagged, but I was also tired, so that was yet another reason we spent a lot of time in bed.

During the time we were apart, we missed several dates together that included Christmas, New Year's, our wedding anniversary, and Natasha's birthday. So, we chose a different holiday to celebrate every day after I got back. During that time, we could almost forget about the fact that there was a war going on or what I had gone to do and what we had to do next. I never told her about the customs and security experience.

On a funny side note, the people in Texas had begun joking about how all the storms and shit was my fault when I was there. Shortly after I left Florida, the place got hit by a tropical storm named Bill. Maybe the people in Texas were right. I hoped not at the time because that would mean the locust and shit were following me. That was not something I wanted to follow me while I was trying to get out of Ukraine. I wrote it off as bad luck and coincidence and kept moving forward.

I had missed the good food in Ukraine, so Natasha and I hit all of our favorite places right away. We began talking of getting out. I knew that Ukraine was her home and it was much easier for me to leave than her. On top of that, it wasn't a bad idea to be in a place where we had family. There is always strength in numbers, especially when people are trying to kill you. Natasha did, however, reassure me that she wanted to go. I told her I would stay in Ukraine if she wanted, but she convinced me she wanted to go even though it meant she couldn't be flying back and forth every time she got homesick.

I took her word for it. However, even though Natasha thought that she had traveled around the world, this trip had no end in sight. Additionally, it would be without tour buses or Russian translators. Once again, I was just going to have to roll the dice, trust her, hope she was as strong as she thought she was, and see what happened. Volodia was an entirely different problem.

As I have already mention, Volodia's father was told by his mother to divorce Natasha when she got pregnant so she wouldn't get anything from the family. However, it was even worse than that. As I heard it, even though Volodia was too young to remember the situation, both his father and Natasha told him that his father was actually his uncle. Sergey was a real pussy and grandmother wanted to protect the family fortune.

Now, don't get me wrong, they were fairly well-off, but it was no major fortune. It was a bunch of rental properties that provided the family $5,000 a month, which was a lot of money in Ukraine. When Volodia's dad, Sergey, who was pro-Russia and pro-Putin, found out about me, he made it very clear he was not going to let his son grow up with an American. If he didn't give his written permission for Volodia to leave the country, we would be left with only a couple of options that I discussed with Natasha. In short, she would have to convince Sergey to let him go, we would have to leave Volodia behind, we had to be even more creative about how to get out of the country, or I had to kill Sergey.

I knew killing Volodia's father would be easy. His movements were not difficult to predict and it was a war. I wouldn't even have to hide the body. Just one quick shot. It would simply be a knife from behind right into the lung, so he couldn't scream, and finish him off by cutting his throat. Even he wouldn't know who did it. I told Natasha to just look the other way and I would handle it. We went for wine.

Natasha thought about it for a while, but in the end, she told me that if Volodia ever found out that I had killed his father he would never forgive me. She didn't want to take that chance. I think she was more concerned about the fact that Volodia would know I wouldn't have done it without her knowing and she was more worried that she would never be forgiven. I kind of agreed with her. Eliminating that option, pardon the pun, we began working on the remaining ones. After all, if Sergey would have died before we left, Natasha was going to be looking at me. I couldn't have just said he must have slipped and fell on a knife.

There are only a few ways out of a country even if it is Ukraine. Those would be air, land, or sea. All of these options were available to us.

The air options were simple. We either had to walk right out the front door of a 2-gate airport, which was extremely bold and stupid by many people's standards, or we needed a private plane. I knew of a military pilot in Turkey who had been a student of mine, so I tossed around the private plane idea for a while over a few beers. There was no way we weren't going to be spotted in an airport with only 2 gates.

The sea option was also viable. All we would have to do is bribe our way on to a boat headed to Turkey and fly out from there. Odesa was the largest Black Sea port, so this didn't seem like a huge task. However, no matter how we got out of Ukraine, we would eventually need a commercial flight, which presented another problem. Natasha couldn't even change planes in America. The ground option was just too long and dangerous given the war, so I didn't try to give this one much thought.

I talked to Natasha about it and checked flights online. It seemed like whether it was Turkey or Ukraine, a commercial flight was going to have to go through Moscow. Given the war and a Russian invasion, I wasn't in a hurry to be the married couple who were formally government, and in fact, two different governments, walking through a Moscow airport holding American and Ukrainian passports. However, at the moment, I was fucked for alternatives.

In the end, I settled on the go big or go home strategy. Once again, I made up my mind that I would not be going to some Ukrainian or Russian jail though. If caught, they were going to have to take me. I didn't like jails in America, so I was certain I wouldn't like them in Ukraine or Russia. We were going to walk right out the front door and go to Moscow.

By this time in the war, Ukraine and Russia were really getting on each other's nerves intentionally. One of the ways they rubbed each other wrong was by restricting flights. This meant it was going to take a little more than an online booking site to get a flight. Natasha immediately went to work identifying a travel agent she thought we could work with. I had a feeling this was another situation that required beer. I should have stuck with that feeling because, as it turned out, I had no idea what I was in for.

When we got to the travel office, I had an American line of thinking going on or at least a Ukrainian one related to the time that we booked our trip to Egypt, but I was nowhere in the ballpark. This wasn't easy. The owner of the place was on the phone with Moscow screaming at people, I was screaming at people, and his girl and Natasha were talking to each other in Russian saying it was a good thing that he and I didn't understand each other because we were both getting pissed and saying really "nice things." We weren't mad at each other, but it was a seriously frustrating situation. During the course of this endeavor, I took Natasha outside. I asked her again if she was sure about doing this, especially since she had made up her mind to leave Volodia behind. She reassured me she wanted to go this way. Three hours later, we finally got the tickets and went for a beer.

Sitting at the bar with tickets in our hands, the seriousness of our situation really began to settle in. Now, we didn't even talk about it. That was a long, slow beer to down and then I had another. Deep down, I knew I could make it and had no other choice now, but Natasha was unproven in this area. Of course, because of my love for her, this was a great concern. All of this was being done just for her. I didn't know how it would turn out.

We still had some things to do to get ready. Natasha had to decide what she needed and could get in two checked bags and one carry-on. I even bought her a new bag just so she could have some more room. However, like a typical woman, when she was supposed to be breaking down her shit and distributing the rest, she went shopping. I shook my head, went to the bar, and decided to leave it alone since things were going to be stressful enough.

On a lighter side, we still had a few moments to have a nice time before we left. We needed to spend these with Volodia and explain to him what was going on. One of those moments was my birthday. I hated the idea of leaving Volodia, but in the end, it wasn't my call. I couldn't do anything about it.

Natasha's beautiful sister, Lena, got me a massive bottle of cognac that came in very handy to help me keep my shit together when it was time to leave. As I have already said before, there were several ways that this exit could have gone wrong and only one way that it could have gone right. At the time, I really didn't like the odds, but I didn't have a choice.

We had scheduled the flight for the middle of the night. In my experience, either the stupid or the extremely tired get that shift, so I thought it would be the best time to try it. I had distributed the weapons I was pretty sure would get me busted when leaving Ukraine or entering Russia. I only packed weapons that I had taken back and forth before, hoping that I would be ok.

It had only been about 6 weeks since I had last entered the country, but I guess my entrance was pretty grand, because I drew a lot of attention leaving, which basically meant I had been flagged, especially being the only cowboy in town. We were expecting it, but I can't help but feel we got a little extra that wasn't deserved. All we could do at the time was push on.

We said goodbye to Natasha's sister and mom at the airport and went to our first hurdle. I had taken a couple of good pulls off of the cognac bottle, so I was fairly certain I was prepared for whatever came next, but wanted to stay loose. We got to the counter and were thinking that things were going smoothly until they saw Moscow on the itinerary. That's when the shit started. Some people fell on top of us and escorted us upstairs for a "private" conversation.

This is where it would start to get dicey and my pucker factor surged up. This was going to be all Natasha's show until somebody drew a weapon. She managed to get us through the interview and I was downright impressed. She once told me that part of her techniques, especially since they were usually guys, was just looking beautiful. This time, obviously worked. I loved her for that. We headed back downstairs for the next stop, which was customs and immigration.

Natasha had been prepared for this one before we ever got to the airport, so she had me go with her mother to do some paperwork before we left. I wasn't quite sure what it was all about, but it had to do with me being registered in Ukraine. As with all former USSR countries, if you were a resident or a citizen, you had to be registered at an actual address. In all-star fashion, she handled that incident, too. I was waiting for her to do everything including taking off her clothes to get it done. Hurdle three, at that point, had been overcome.

As I was speaking to the customs guy, I looked around for Natasha and realized I was surrounded by security. Shit was about to get real at this point. They went to grab me, but I wasn't going to have any of that. I didn't know where Natasha was, but she had been there just a second ago, so I knew she wasn't far. I was beating myself up for losing focus enough to lose her, but I knew that most likely after three checks already, we were headed to the same place.

Normally, interrogation would take place in a closed area where sound would not be much of a problem. That was an advantage for them, but also for me. They ushered me to a room where I found Natasha parked in a chair in a corner. Two armed men had emptied my bag and had spread my military-grade weapons out on the table. Through various reflections in the room of the door behind me, I realized they had not manned the door. That was a mistake.

They began to ask the real questions now. Was I military? Was I a spy? What country was I affiliated with? I went into interrogation mode and gave them nothing more than they already knew. I could see Natasha was worried, but I gave her a nod that things would be fine and played the stupid tourist. I made up my mind that one way or another, we would be leaving that room as free people.

At that moment, I was left with just a few options that were immediately obvious. The first one was to be taken, which I was not going to let happen. The second was to kill them and either try to make the plane before their bodies were found or head out the front door back through all the other checkpoints and execute plan B on the sea. The other option was, of course, to end up dead.

All of a sudden, I caught a look in one of the guy's eyes. He kind of had this "let's make a deal" look on his face. These guys were not paid very well and they certainly didn't get the best gear. I went for it. I decided to offer up some of the gear to keep the rest and avoid the other bad options. They went for it. We divvied it up and Natasha and I were on our way.

We got on the plane and I got a fuck load more drinks before we were to arrive in Moscow. I figured getting out of Ukraine would be the easy part. That hadn't gone too well since I almost had to kill a few people. Of course, as figured, when we got to Moscow, there were more problems.

The good news was we were past initial security and now we just needed to get through the transit zone in Moscow. The bad news was it was Russia. We were detained again and told that we could not go to the plane. Natasha was back on point.

While she quickly explained to me that they were not saying nice things about us, I kept watching for security. As Natasha argued with people in Russian, I kept an eye on the plane, too. Eventually, through Natasha's skills, she got us on the plane. After we boarded, we got hammered, landed in Madrid, drank some more, and poured ourselves back on the plane for the long leg to Ecuador. Next, we were going to have to get into another country, since we had gotten out of the one we were worried about. However, we were on one-way tickets. Which, as I have mentioned, was usually instant deportation back to where you came from.

Chapter 14: In Ecuador

On the plane, I got one more drink, kissed Natasha and settled in for a nap. We still weren't through, but I was hoping that the worst was over. In all fairness, I had left Natasha out of the loop on a few issues and didn't want to have to explain that later. I slept pretty well, but did get up to watch a movie as the jet lag started to set in.

As we neared Ecuador, the announcements started, and they began handing out entry forms. I started to school Natasha. As a lawyer, she figured she knew this shit, but really didn't understand our current situation. After some back and forth, I did manage to get her to listen to me and she filled out the forms like I told her to while I got another drink.

She may have been great as a lawyer, but I really didn't know. What I did know is that she was really good at talking too much. That was not what we wanted to be doing while trying to enter a country from Ukraine on one-way tickets. When we got on the ground, I simply told her to keep her mouth shut and if anybody asked her a direct question to only respond in Russian.

We got off the plane, found our shit, and made our way to customs. If you haven't figured it out from reading the book this far, I had traveled a lot and that probably turned out to be a good thing. The customs officer was so caught up in my passport, since it barely had any free pages left, that he never did ask questions such as the reason for our visit or how long we were staying. Lucky for us, he never even looked at our tickets for a return date, either.

He started with Natasha first. She didn't have a single stamp in her passport, because she had just changed her name and got a new one. He simply looked at me and asked if my wife was Ukrainian. I said, "Yes." He then started through my passport. He asked, "You are a US citizen?" I replied, "Yes." He asked, "You are a Ukrainian resident?" I replied, "Yes." He asked, "You have a 5-year Saudi Arabian visa?" I replied, "Yes." He asked, "You spend a lot of time in Turkey?" I replied, "Yes." Then he asked, "Where do you live?" I simply said, "That is complicated." He said, "Welcome to Ecuador," stamped our passports and we moved on.

Having a bit of a lighter load in the weapons department this time, we went through the security check without any problems at all. We had gotten in late in the afternoon, but it wasn't dark yet. We made our way to the taxis and loaded up our stuff.

On the hour ride into the mountains to Quito and to our hotel, Natasha and I were both a little sick from the altitude. It had started to affect me before we even landed. However, we made it to the hotel and I contacted the lawyer. He said since we had gotten through customs, he would show up at the hotel the next day and take care of everything else. We got a drink, but I could barely keep it down. I ended up puking for two days because of the altitude sickness.

We met with the lawyer the next day, listened to what he had to say, and got him some more money. His job was about to begin, now that we had arrived. Afterward, we set out on a search to take care of communication problems since we were in a new country. We got cell phones taken care of and then started to get the lay of the land. We had no idea where in the city we wanted to live. The lawyer was useful here, too.

Natasha pretty much went into holiday mode and seemed to forget the whole reason we were there. I entertained it for a little bit because we probably both needed some downtime, but this was no vacation and we had shit to do. She was running around trying to hit all the tourist spots in town in a matter of days. I think we must have looked at twenty churches at least. The upside was they were all in different places of the city so I could try to figure out where the best area would be for us while letting her wander around.

While we were cruising the city, it became obvious that the south really sucked. Luckily, I had gotten us a hotel in the middle of the city. North was where we wanted to be and it was by the embassies and UN building, not to mention, the Olympic Stadium and shopping. There was also a park for Volodia if we could ever get him out of Ukraine. We thought he should have great kids to play with at school that would also be foreigners. Regardless, I told Natasha she was going to have to get out of tourist mode soon and focus on a place to live because we couldn't stay in the hotel forever. I found us an English-speaking real estate agent.

We didn't look at too many places because we needed to get out of the hotel and tourist price zone. After about the fourth place we looked at, we decided to take it. It was fully furnished with brand new appliances and linens for the bed, was across from a park on one side and overlooked the Olympic Stadium on the other. We could watch football games without ever leaving the building. It also had 24/7 security, a pool, a sauna, and conference rooms we could use at no extra charge and was only $650 a month. I thought that was pretty damn good at the time, but I later found out we could get something much cheaper.

We still needed a maid and to set up utility accounts and internet, but I enlisted the realtor and the security guys in the building for this effort. In short order, we were up and running. The apartment was even pretty close to the lawyer. He stayed on top of everything and it looked like we would end up with Permanent Residency very quickly without having to have Temporary Residency first, which is what many countries in the world would do, including Ukraine.

We started to settle into our new environment, which really tested Natasha. She wasn't much for speaking to strangers, but I knew if she didn't have any friends it was going to be much harder for her. On top of all of that, she didn't work and was going to need something to do so that I didn't have to deal with entertaining her all day long. I was writing my first book and entertaining her just wasn't something I had time for.

So, on top of learning to deal with the Spanish-speaking maid who also cooked meals for us, since Natasha would burn water, I also got her a gym membership and private tennis lessons. I took her back and forth to the grocery store and clothing stores until I knew she would be safe doing it by herself. Then I began to settle into my own schedule. Of course, girls still like to go out and dance or just eat somewhere new from time-to-time, so we also began exploring other areas of the city. Natasha had even learned of Russian groups within the city, but she never went to any meetings or talked to anyone.

She still had the notebook computer I had gotten when mine went missing during the Malaysia trip. She was able to talk to her family back home, which made her feel good. I was happy for any peace she could find in life with the transition, but it was becoming clear that something was going wrong with my body. It seemed like the more I settled into daily life and slowed down, the more I was in pain.

I didn't really think about it at first and blew it off for the most part since we had some new issues to address. Apparently, we had arrived just months after the volcano outside our fucking window had gone active. With this new situation, we got to experience ash, earthquakes, and motherfucking wildfires that threatened our building. I was beginning to think about the locusts again.

The US Embassy called an emergency meeting for Americans in the area and I went because, of course, I wanted to know what was going on. Additionally, a name that I had been tracking on the expat sites on the internet would be there and I figured he would be a good source of information for many things. I needed to get tied in and connected. He showed up and put together an after-meeting event that we were invited to. I swung by the apartment and got Natasha, hoping she would get used to meeting new people or at least be happy about going out.

Ultimately, all they had to say about the volcano was that it was active and people were watching it. I wasn't really sure what that meant because it was not like they were going to keep it from erupting. They talked about what supplies to have on hand, what to do about ash, and be prepared to run.

That said, it was nice to meet people in the area, and being new to town it helped us settle in, or at least it helped me. Natasha barely engaged with anyone, but that was no surprise. I got many questions answered and established important relationships. I could have probably established some of those relationships virtually, but it is always better when you can have a beer with a person and be somebody that they will later think about. These moments are sometimes rare and you have to take advantage of them when you can.

As if things hadn't been shitty enough, we also arrived at a time when many people in the country were upset about the president. There were many demonstrations. Natasha obviously began to question my sanity and judgment, even though she probably had before, because we had just left a situation like that, but, at least, we hadn't had to deal with volcanos in Ukraine. I guess I kind of understood her point. On the surface, it was not looking good initially, but I decided to stay the course.

I didn't really understand the protests, but people were fickle in general. If anything went wrong with a country, they were going to blame the man at the helm. In this way, it was no different than being the captain of a sinking ship or the project manager of a project in trouble. This president had been in office for almost 10 years and even stepped down once to be re-elected again as I heard it. From what I understood at the time, they had even changed the constitution so he could run again. I had no idea why the people were pissed now.

Every day I was working on about ten things at once to get us stable. I was in constant contact with our lawyer about residency. I was going from restaurant to restaurant and bar to bar to meet people and figure out what was going on in our new world. I was trying to figure out how we could maintain our survival in a new country where we had no real connections. It was scary when you took the time to think about it, but not disastrous. We had made it to Ecuador with about $50,000 left. We weren't in a desperate situation, but it wasn't enough to retire on.

The more I looked at business opportunities, I knew I didn't want to have employees. All the employment laws were about protecting the employees, which was understandable. I was more concerned about something I had heard. What I was hearing was that legal representatives of companies were being arrested at the airport as they tried to fly out because of some kind of labor dispute with their company. This was not something I wanted to deal with. I continued to focus on internet-based businesses.

I did consider "hiring" people as "contractors" for internet business ideas and tossed this around with some lawyers, but ultimately that was a stupid idea also because the people of Ecuador didn't care about technology. They were descendants of Incans with cows and shit. The last thing they wanted to see was a fucking computer. I kind of admired them for that. The president had tried to make the country the technology center of South America by setting up a university that citizens could go to for free. He couldn't fill the fucking place! So, they started to bring in foreign tech labor. I thought that could be an option.

As the days passed, we got closer and closer to being legal in the country on a permanent basis. We still needed a fucking income, but at least we would have a country. After the way America had treated us, I was ok with at least having a country again. It is amazing how you begin to feel in these situations. Things you thought were only for TV and movies had now become your life. It is surreal, to say the least.

Office visit after office visit with our lawyer, we got closer and closer to our residency. From what I could see, he was moving pretty quick. I didn't know how quickly until I began to see how shocked other people were when discussing the situation. They also started asking for my lawyer's name and contact info. At least it appeared that I was making some of the right contacts in the country, but I was fairly certain I was making some of the wrong ones, too. The trick is to know the difference.

All in all, day after day, my body got worse and worse. I was shutting down whether I wanted to or not, but I had to secure Natasha first. There was no way in hell she was going to know what to do on her own and it would scare the hell out of her when I finally went down. She just wasn't that kind of girl. In that way, I kind of missed the cowgirls of Texas, but that wasn't going to change my situation in Ecuador. I had to just keep finding a way to put one foot in front of the other every day.

The day finally came when I woke up, put my feet on the ground to get out of bed, and fell flat on my face. Once again, I had pushed it too far. Natasha was at the gym, so I was flat ass fucked. I screamed and hollered in pain as I made my way to a desk and pulled myself up. After that, I screamed and hollered in pain some more holding on to walls and furniture while making my way to the phone to call for security. Now, I don't know if you have ever tried it, but it is extremely difficult to scream for help in a foreign language while in a shitload of pain.

It didn't take security too long to get to the apartment. Luckily, they got there about the same time as Natasha. They probably thought the crazy gringo was drunk again. Natasha probably thought the same until it became obvious I was actually fucked. Between Natasha and security, there were three or four of them, and they got me to my feet and into a cab.

Now, for the people that know me, I didn't go to hospitals unless I was dying. When I was in foreign countries without citizenship and no insurance, I better be at death's door. We got in the taxi and I had the guy take me to an American chiropractor I knew about. Of course, once again, I was too much John Wayne and cockroach. The doctor took one look at me and said "Hey, listen, hard-ass, you don't need a chiropractor. You need an emergency room. I am arranging it now."

He gave me a shot of something to make me happy, some travel needles to go, and put me in a taxi. Just like he said, the hospital was ready for me. When I arrived, security came out and got me out of the taxi and into a wheelchair. There were no questions, no forms, and no wait. They wheeled me right to the back of the aisle where the doctor was waiting and promptly stuck a needle in my arm to help get me going again.

While he wired me up for an IV, he asked questions and I gave him the rundown as I knew it. After about an hour and three bags of morphine in drip form, I was high as shit and ready to walk, run, or fly.

He, of course, told me I needed some serious follow-up and I took it under advisement, but all in all, I was feeling better. Fuck that! It was time for a beer.

I got settled back into the bed that night, juiced myself up with the needles, and did the whole John Wayne hot tub of water and alcohol thing until it became obvious, while I was crawling around on the floor, that the process wasn't working. I juiced myself up again and went to the pharmacy to get a back brace, ACE bandages, and shit because, for some reason, guys pretty much figure that everything can be fixed with duct tape. Apparently, I stuck out like a sore thumb. As I was coming down the escalator, I ran into another American doctor that convinced me it was time for X-rays and to meet with him in his office.

I was concerned about the fact that I stood out so much since I hadn't been in town that long. However, I was in too much pain to give a shit, got the X-rays, and went to see the doc. He got me on a table, whipped out his protractor and shit, and went to work on the X-rays while I laid there. After that, he came to me with "the" look on his face. I knew "the" look. It was the one that always told you that you were fucked. I wasn't feeling great physically or emotionally, at that moment, and I didn't have a beer, so I felt even worse.

I had never wanted to be the guy lying on his death bed regretting all the things he didn't do or the hard jobs he didn't take. Ukraine had finally taken its toll. He basically told me the injuries to my back were "hopefully manageable," but there was a good chance that I could end up with plates in my spine in addition to all the other plates in my body. I just acknowledged it and we moved the fuck on, setting up three appointments a week, including some more electroshock therapy.

With the help of the doc, I began to get better and started walking again. However, after one particular visit where he tweaked me out, I may have been feeling too good. So, I went to the Irish bar for a while. I knew I had some time because Natasha was at the gym and had her tennis lesson after that. Quite frankly, I hadn't been walking around for a long time and wanted to get the fuck out for a while.

Eventually, I left the bar and started making my way home. Getting out of the taxi at the next bar convinced me that I had started thinking I was better way too early. Once again, I stepped down and fell flat on my face. Only this time it was in the middle of a main street with heavy traffic.

Lucky for me, the visits to bars and restaurants had paid off. I managed to crawl my ass out of the street without being hit by a car and made it into the restaurant where they knew me, helped me get to my feet, and got me a beer. After that, I was able to call another new contact to come and take me home. He showed up with some local "medicinal" items to make sure I would be out of pain when he left me.

I got home, but was laid up again and went through another phase of the healing process. It was clear that I was going to have to take things much slower than I wanted to or was used to. I guess in that respect, not only the war and my past life were catching up with me, but also age. To add insult to injury, I had asked the doc about going back to my old life. I had heard about a contract in Beirut and was considering it. He just looked at me like I was stupid. I realized I was going to have to re-invent myself even though I didn't really want to or know how to. I was back to one step at a time since everything was catching up with me.

In time, the doc got me straightened out, sort of speak. I started looking like my old self and performing a bit more normally. This may sound like there was too much going on to deal with residency, but that continued to move forward. The day finally came when our lawyer called and said we were good to go and needed to get our pictures taken.

This was good news because it meant that we were finally safe, even if we didn't really have a survival plan yet. We met our lawyer at the appropriate government office, had our photos taken, and left with our national ID cards. Additionally, we also had the right visas in our passports, so we were looking good. We were now officially permanent residents. It was time for us all to go out and get beers together.

This put even more pressure on me to work out the revenue plan. I began asking the lawyer a completely different set of questions to determine legal options and costs. Like many lawyers, he had his finger on the pulse of the community and did his best to enlighten me. When everything was considered, I really didn't want to be filing foreign tax returns in America, especially since the country had fucked us over. I mean, if I knew my fucking tax dollars were never going to fucking help us, then why should I have to give them a fucking dime.

I was moving around better, but the doc made it clear that the shit I used to do was in the past. I started to think about the training option a little differently. I had noticed some things in Ecuador. For the most part, the military, police, and people responsible for security were stupid. I saw this as an opportunity, but getting it past Natasha would pose an issue. Well, that made it just another day at that point.

Conclusion

We had come to the end of this part of our journey and adventure. There were still questions to be answered and I was sure that there would be more challenges in the future, but it was the life that we had chosen as soon as we decided to board a plane on one-way tickets to a place we had never been before. Another person will never know what it feels like until they make a decision to leave everything and everyone behind based solely on the fact that they had no other option.

I sat at the bar with another beer reflecting on what had happened in just a few years. At the beginning of 2013, I didn't even know of Natasha. By July of 2015, however, we were married, had sold everything I owned, successfully avoided the Ukrainian mafia, armed Russian soldiers who invaded the country, armed Ukrainian rebels, the Odesa massacre, and had moved to a new country on one-way tickets and established permanent residency. The locusts never came, but I would keep looking for them.

It was a lot to take in and I was happy for every bit of the training I had received over the years of my life. It had proven to keep me alive. I was now in a position where I had done things that the rest of the world couldn't even imagine. More importantly to me, I had made the hard decisions, been tested under fire, and came out well, having proved myself. Other people in this story were not as lucky.

I still had no plan for the future except for maybe lunch. Hell, at that point in time, I would have taken a job managing one of the expat bars or something. I did respect all of the Venezuelans pouring into the area to find work and escape their failing country. These people were tough and had seen many things.

They had nothing and I could definitely associate with them. They went from being lawyers, vice presidents of banks, civil engineers, and other professionals to bartenders and prostitutes. I was receiving a real lesson in how a country collapses and what it looks like when it does. Between Venezuela and Ukraine, I couldn't figure out what the fuck many Americans were thinking about or why they were crying out for so-called Democratic Socialism. It was just socialism by another name.

Some of these people, of course, became very desperate and took to a life of crime to survive. Most of them, however, had too much character for that. They would prefer to sell their bodies for sex rather than steal from someone else. I assumed that they understood that mistreating others by stealing from them was not the way to recover from their tragedy.

Depending on how much money they still had or could get, we lived in the same buildings with them and, regardless, we lived around them.

Some of them were lucky enough to be able to afford a bus ticket from Venezuela to Ecuador and were able to leave with a bag of clothes and a bag of food. Others were not so lucky. They had to walk out of Venezuela across Columbia and into the mountains of Ecuador 9,400 feet high.

Regardless, they would do some of the most amazing things that made you wonder about Americans living in the USA. For example, many times they would be cooking and would come over to ask if we wanted anything to eat. An entire group of desperate people that had been starving was concerned about whether or not their neighbors were hungry. It was both touching and humbling. I didn't want to take their food knowing of their hardships, but I certainly didn't want to insult them by refusing such an incredible gesture.

Natasha and I had been through many things with our experience and many understood a part of that. I guess it was something that bound us. Common hardships and an uncertain future can often do this and we were all looking at it. Since we were able to see so many people around us with the same problems, it was kind of hard to worry about our problems even though they still needed to be resolved. Some people were in better shape than others and some were in worse shape, but we were all certainly in the same boat.

I hadn't worked on my Spanish since high school. In that respect, Natasha and I were finally on even footing because she barely spoke any either. It frustrated both of us, but more so Natasha. I mean, hell, I had been living in Ukraine and barely spoke Russian. I tried to tell Natasha that it had been part of my journey. It would be better if she tried to embrace it and not complain.

Since our Spanish sucked, I didn't really read newspapers or watch TV. When I left the apartment to find a place to work on my book, a TV was normally on with someone who usually spoke English. It didn't take much to realize that something was going on. There was the usual shit about Julian Assange and Edward Snowden who had both tried to get to Ecuador and didn't make it, but it also appeared that the president was on his last legs politically.

I was a little depressed that we were getting there while the president was on his way out. In my opinion, he had done many things to clean up Ecuador, including throwing out the US military and the CIA. What used to be a place with CIA kill squads running around was now a safe zone for just about everyone. Generally, as long as you didn't create problems in Ecuador, they really didn't fucking care what you did. This was one of the many reasons that Assange and Snowden had been headed to Ecuador. It was also why it had been my escape plan for more than a decade.

I continued to make friends. The more I felt comfortable with them and their story, the more I would tell them about our story. However, for the most part, I never used my name. Our lawyer, of course, knew most about our story, but that was a necessity. The day finally came, however, when I received a phone call that left me with mixed emotions, and not knowing whether I was lucky or unlucky.

On the phone was a contact that I had made that had ties to the government of Ecuador. In fact, his uncle had been a Nazi that escaped and made his way to South America, like so many others after WWII. His father and the brother of said escaped Nazi, had made great sums of money selling arms to various South American governments and militaries. Some deals didn't go so well, so he didn't leave the house often since there were contracts out to kill him. After being in the country for a while and meeting his son, he became aware of me and we spoke on the phone a few times.

This particular call came at a very bad time. Natasha was standing right next to me at home. On the other end of the line was my contact who was acting as a middleman and interpreter with the top-ranking Military Intelligence General in the country who was on yet a third phone. I said, "Hello General." Even with Natasha's broken English, there was no covering up that I had just been contacted again about a life that I had promised Natasha to leave behind. We were still facing an uncertain future in a country new to us. Natasha was, of course, beginning to wonder why we had ever left. However, what is life without challenges? Natasha knew who she had married. She just shook her head while looking at me.

About the Author

William H. Volpé III, PMP is the founder of The Volpé Consortium, Inc., the author of the bestselling business book Guide to Project Management Life Cycles, the creator of The Volpé Consortium's proprietary project management process, the publisher of Project Victories, and the developer of a three-day project auditing seminar.

Mr. Volpé is an expert project management consultant having more than 30 years of experience in project management. Also, he was a Master Business Analyst and Worldwide Manufacturing Program Manager at Hewlett Packard as well as the PMO Manager for the Ministry of Interior of Saudi Arabia.

His experience includes several multi-million dollar international projects and he holds an Associate Degree in Computer Engineering Technology, a Bachelor's Degree in Electronic Engineering, a Bachelor's Degree in Management, and a PMP (Project Management Professional) certification.

Additionally, he holds a Master Certificate in Project Management, an Advanced Master Certificate in Project Management, and a Professional Business Analyst Certificate, all from The George Washington University School of Business. Other certifications include an ISCET

Journeyman in Computers, an ISCET Journeyman in Industrial Controls, an ISCET Journeyman in Communications, and an ISCET Radiotelephone License.

As a talented and experienced project management consultant, Mr. Volpé teaches classes and consults in project management internationally. Classes include Project Auditing, Project Risk Management, Managing Projects, and Project Scheduling and Cost Control. He has taught for such companies as Verizon Wireless, Dow Chemical, Shell Oil, Halliburton, the Internal Revenue Service / the United States Department of the Treasury, and the National Geospatial Intelligence Agency.

Mr. Volpé began his project management career as a U.S. government contractor under Secret Clearance supporting Operation Just Cause, Operation Desert Storm, and Operation Restore Hope. He worked at Hewlett Packard in the NonStop Enterprise division and the Business Critical Servers Business Management group for more than five years; was a pre-merger Compaq employee; and completed several new product introduction projects for fault-tolerant, mission-critical customers such as the Pentagon and the world stock markets. Mr. Volpé has extensive international project experience and his teams have included members from Germany, England, Mexico, Australia, Scotland, Taiwan, and several other countries.

Mr. Volpé still consults on project management today and can be contacted at contact@thevolpeconsortium.com.

Made in United States
North Haven, CT
05 November 2026

21758668R60146